THE MAN WHO FENCED THE WEST

By

Glen Coleman

Illustrated by Siegfried Reinhardt
Edited by Shirley Schoonover

Publishing Co.: Glen Coleman Company

ISBN 0-9614346-0-0

Acknowledgements

Siegfried Reinhardt

World renowned artist, Siegfried Reinhardt, of Kirkwood, Missouri, has provided the beautifully detailed, black and white illustrations that appear throughout the book. The late Mr. Reinhardt has paintings in the Whitney Collection and the Vatican.

Shirley Schoonover

The book was edited by Shirley Schoonover, who has won an O'Henry Award for her short stories. She has also written a novel that was runner up for a Pulitzer Prize.

I would also like to thank Lou Liebermann for indexing the manuscript, Ralph Castle and Warren Loveridge of Laclede Steel and Albert Moorehead of Salt Lake City, Utah.

Information in this book was obtained from conversations with E. A. Staman, Paul Sippel, Rome Blair, Felix Guelfo and Pat Meador, all of Shreveport, Louisiana; Everett Nelms of Montgomery, Louisiana; David Young, Crawford Young and L. F. Young of Campti, Louisiana; Leonard Huber of New Orleans, Louisiana; A. F. Nickelsen of the Placid Oil Company; Fred Clifford of St. Louis, Missouri; Harold Vollbath of Kansas City, Missouri; Robert Weaver of Chicago, Illinois; Gordon Dippel of Bayou Sara Louisiana; Jas Fair of Austin, Texas; Art. H. Brockelmeyer of Kirkwood, Missouri; Ralph Myers, a famous engineer of Chesterfield, Missouri and Dorothy Drain Coleman of Mesa, Arizona.

Organizations that contributed to the book were:

- St. Louis Public Library
- St. Louis County Public Library
- Museum of Transport, St. Louis, Missouri
- Tulane University Library of New Orleans, Louisiana
- New Orleans Public Library
- Louisiana State University of Natchitoches, Louisiana
- Chicago Public Library
- Chicago Historical Society
- Illinois Historical Society
- Missouri Historical Society
- Mercantile Library of St. Louis
- Ellwood House of Dekalb, Illinois
- Kirkwood Public Library of Kirkwood, Missouri
- Gonzales, Louisiana Public Library
- Coushatta, Louisiana Public Library
- Southern Methodist University Library of Dallas, Texas

TABLE OF CONTENTS

The Man Who Fenced the West

ONE

1866
The Immigrant

The eighteen year old William Edenborn, newly arrived to America, looked anxiously over the ship's rail to the scene below. He had arrived earlier at the port of New York City, only to be hustled back aboard the ship to disembark here at Castle Rock. "Bedlam," he must have thought as he watched the confused, jostling mass of people below him.

Apprehensive, William glanced again at the book he'd been given aboard ship. He reread the pages, and the promises the book made: he would be escorted by government officials through the proper channels where he would receive a free medical examination; he would be processed according to nationality and language; his name, country of origin, his destination would be recorded; that information would be forwarded to the information department or a registered agent of one of the Transcontinental Railroad companies. Then he would obtain a ticket, check his baggage, transfer to a railroad station where he would be led to, and board, the correct train for the great American west. The book even assured William that the insane and the criminals would be weeded out of the groups of immigrants. And, the book promised, everything would go smoothly because of the efficient, friendly officials here on Castle Rock.

Trusting the book, William stepped ashore and was immediately buffeted by frightened, confused strangers who spoke languages he couldn't understand; William was pummeled, nudged, elbowed by the crowds that surged around him. Someone stuck a fist into his ribs and he was herded into a slow moving line . . . "like animals," he thought.

William saw an official looking man, an immigration agent, standing behind a table. William tried to ask him where he should go, but the man yelled at him, voice harsh, words alien, and shoved William back into the line. William called out again, his English heavily German accented, "Please, would you tell me . . ."

But the immigration agent glared at William, yelled something at him, then turned his back to William.

And the mob of frightened people shoved William along, away from the officials, away from the answers he wanted. Dazed by the hysteria around him, William allowed himself to be crowded against a wall where he decided to rest for a moment, to think about what he should do next. As he looked around at the milling strangers, he yearned for the familiarity of his home in Altena, the village in Westphalia's Ruhr Valley.

Altena, where wiremaking was a tradition, where families took their names from the wire: Drahtmeyer (wiremaker), Cable (Two Wires), Cabell, and Koebel. These names had become famous centuries earlier for the crafting of chain mail armor for the medieval knights, and these same families had developed the "Sull Coat" which was a way of surfacing the wire before it was drawn through the wire die. The "Sull Coat" made wire drawing easier, but would do minimal damage to the surfaces of the wire and the die, the tapered hole through which the wire was drawn. The "Sull Coat" was such an important process in wiremaking, that at one time the Baron of Altena had kept it a jealously guarded secret. He had been so fearful of losing the process that he refused, for a time, to allow people to move into Altena; and once they had lived in Altena, he would not allow them to move away for fear that they would take his "Sull Coat" secret with them.

William had been born in 1848 in Plettenburg, a small village near Altena. His mother, Antoinette, was of the respected Hessmer family; his father, Jacob, owned a small greenhouse and farm where he raised plants and seeds to sell to the neighboring farmers. William's parents died when he was thirteen, so he moved to Altena where he learned wiremaking. Industrious and well-mannered, William was treated kindly by the folk of Altena, yet he never felt completely at home there.

Aware that most of Altena's wire production was being sold to railroads for telegraph use, William concentrated on expanding his craft as a wiremaker. A brilliant student, he financed his education at the gymnasium while he worked at wiremaking. He read everything he could find, including a collection of letters by Gottfried Duden called *Report about a Journey to the Western State of North America.* He read, "The American farmer must never be thought of as a European peasant transplanted. In dress and deportment, city and country folk are alike. All professions are equal in rank. No one feels reluctance in doing manual labor. No prejudice prevents even a doctor, a judge or a colonel from tilling his own land. When in Europe a rural community attempts to take on the ways of the city there is always a tremendous lot of harm done, because the evil rather than the good things are emphasized. The best remedy is to have the young people of such communities emigrate to a country like America, where the joy of labor holds in check enervating excesses, and the abundance of everything makes poverty with its accompanying acts practically an impossibility."

Reading letters like Duden's, William began to think about America. As a wiremaker in Altena, he knew his future would be secure; but it would also be narrow. The ruling families of Westphalia and Prussia did not intend for the trades people, the craftsmen to ever have more than a living. There was no room or freedom to move up. And William, even as a youngster, wanted that freedom. He read: "The farmer comes in touch with the most intellectual men of the nation, especially when his vote is solicited before the election." Duden's letters continued, "The officeseeker visits with him, personally, or he hears him discuss the issues of the day at horse races, the mustering of the militia or other such gatherings. The different views of the various candidates is put before the people so that no one can help getting something. It is surprising what a change comes to people in one single generation—from the position of a German peasant or tradesman to an American citizen."

This was heady reading, indeed. These were turbulent times for Westphalia. In 1848, the year that William had been born, abortive revolutions had attempted to curb the powers of the nobility. Then, twelve years later, Otto Von Bismarck, the Iron Chancellor, was recalled from St. Petersburg, Russia where he had served as emissary, to assist the weakling King of Prussia. Bismarck had instituted the draft, and had extended the period of military service. A far cry from: ". . . when his vote is solicited before the election."

William continued to read Duden's letters: "How often have I thought of the poor of Germany. What prosperity and affluence could a few diligent hands prepare here for whole families, whose condition in the Fatherland is wholly inconceivable to the planters who were born in

America. Along the Missouri there is still room for millions of beautiful plantations, not to speak of the other rivers at all. The great fertility of the soil, its enormous expanse, the mild climate, the splendid waterways, the absolute freedom of intercourse in a territory of many thousand square miles, the absolute safety of person and property, the very low rate of taxation, these are the things which must be regarded as the real foundation of the fortunate position of the Americans."

Duden's thirty-first letter continued: "In what other country on earth are all these combined? If it were desired to embellish the picture still further, it would only be necessary to remind the reader of the rich forests, the superabundance of coal, salt, iron, lead, copper, saltpeter and

1860 Wire Town of Germany

other minerals; of the inclination of almost all the inhabitants to utilize the natural advantages; of the river navigation which even now begins to flourish; and finally of the absence of all European prejudices in regard to rank, to trade, and to physical work."

Earlier, Duden had written, "By careful reflection I had come to the conclusion that most of the evils from which the inhabitants of Europe and especially of Germany are suffering, arise from the effects of an excess of population . . . I recognized at the same time, that excess of population distorts the social condition in a lamentable manner, and that all the endeavors of the intellectual forces of society must necessarily end in making the state nothing but a great prison-house."

William couldn't help but think of that term, 'a great prison-house' when Prussia used the railroads to defeat Austria in the Six Weeks War and annexed Westphalia.

America seemed to beckon to him when Duden's thirty-first letter added: "German immigrants are always welcome here, and as soon as they have entered upon the new continent they are regarded equal to the citizens. There are but few trades which do not prosper here."

William's first job in the New World was as a mechanic in a small wire mill in Frankstown, PA, a suburb of Pittsburgh. Those first few months in Frankstown were lonely. Although he had resolved to make his life in America, William found himself homesick and a bit sad. "But," he told himself. "You can make America home, too." Frugal and shy, he spent much of his free time alone, reading the newspapers to study the language and to learn as much as he could about his new country. He read about the owner of the mill he worked for, a Mr. Oliver who was best of friends with an Andrew Carnegie. The newspapers called Andrew Carnegie the richest man in Pittsburgh. They called him a "war profiteer." From the newspapers, William learned that Mr. Carnegie had invested huge sums of money in oil and steel and that Mr. Carnegie had taken a steelmaking process by a French inventor, Henry Bessemer, and was making a purer, better steel with it. William read that England had produced a harder cannon ball with the Bessemer process for the Crimean War.

Because he was such a hard worker, and despite his shyness, William's workmates liked him and teased him about his good looks and the way the young women "gave the eye". His fellow workers called him 'Bill' and as those early weeks passed, he began to feel at home in the mill.

After a year in Frankstown, Bill took a train to Cincinatti, where he worked for a few months as a wire mechanic at the Globe Wire Works. During this time he held a second job at night: he had learned early in his life to be frugal and to save every spare penny. This frugality would serve him well in the years to come.

TWO

1869
St. Louis

But Cincinnatti was not the great American West that Bill had read about in Duden's book. He moved to St. Louis, the Gateway to the West, and found a job running a machine that stamped out metal kitchenware. He worked for F.G. Niedringhaus who owned the St. Louis Stamping Company. The company was large, occupying an entire city block in north St. Louis at 2nd and Cass streets. Later F.G. was to lay out Granite City to give St. Louis its first basic steel company.

Much as he liked his work producing the kitchenware (with its special coating this is known as Graniteware) Bill took a night job. He became janitor and watchman at a hardware store run by Frank Ludlow and Oliver Saylor. Bill told Ludlow and Saylor that he knew how to make wire. Since all wire was being imported from Cincinatti at that time, both Ludlow and Saylor were excited by the prospect of manufacturing their own wire. With the aid of Alfred Clifford, the secretary of their company, they costed out the project and decided that it was feasible.

Because of his own frugality, Bill had substantial savings of his own to invest, so after long hours of planning and work, Bill personally drew the first strand of wire west of Cincinatti! Soon after that, Bill left the St. Louis Stamping Company to take on the much more responsible position of wiremaker and stockholder in the Ludlow-Saylor Company.

He didn't stop to think about it, but William Edenborn had begun to make his way in America!

These were exciting years in St. Louis. By this time, 1869, St. Louis was third only to New York City and Philadelphia in manufacturing. Within the next two years, the Eads' Bridge would give the city rail access to the West, making St. Louis truly the Gateway to the West.

Bill enjoyed St. Louis. He joined the Mercantile Library where he studied books and articles about the craft of wiremaking. He joined the Salem Methodist Church, located at 15th and Washington (now Cole) in north St. Louis, where the Reverend Pfaf was the minister. The Salem Methodist Church, one of the oldest in the area, had separated from the Southern Methodists in 1840 over the slavery issue. Bill sang in the choir and met other young people that soon formed his social circle. With his friends, he drank beer at Schnaider's Beer Garden and Brewery near Lafayette Park and at Tony Faust's restaurant. Still frugal, Bill walked everywhere in St. Louis and enjoyed an active, although not frivolous, social life.

Bill met Captain James B. Eads, the designer and builder of the Eads' Bridge. The two men became friends and Eads described for Bill how the bridge had been built. Eads had learned about the movement of water on river bottoms from all the salvaging he had done along the Mississippi, he told Bill. When he had begun the design for the Eads' Bridge, he had remembered a bridge at Coblenz, Germany, and had chosen the upright arch design because it had twice the load carrying capacity of the suspension or truss design. Eads had also chosen the Washington Street location because the Mississippi was narrowest there. With his two assistants, Henry Flad and Charles Pfeiffer, Eads believed that he knew the Mississippi better than most engineers. He had even had the design computations checked by the distinguished mathematician, Chancellor William Chauvenet of Washington University. It was this exacting attention to detail that

1873 Laclede's Landing

guaranteed that Eads' Bridge would be successful. Bill became an admirer of Captain Eads and studied Eads' plans carefully, already aware that he, too, would design and build bridges.

During this period of their friendship Eads told Bill of how he had studied underwater channel pier construction. Eads described how compressed air could be forced into an airtight cylinder at the bottom of the excavation. The sandhogs (workers) would shovel sand into the suction pumps which would discharge the sand at the surface. Eads had also discovered the cause of the Caisson disease (the bends) that had made his men ill; and he had increased the decompression time from five minutes to two hours, thus curing the disease.

But bridges and wiremaking weren't all that occupied Bill's attention. One afternoon he and Alfred and Frank Ludlow went down to the levee to watch the finish of the boat races. The Lee and the Natchez had completed the 1200 race from New Orleans to St. Louis; the banks of the river had echoed with cheers and fireworks. This race was such an event that entire communities had closed down, schools and stores had emptied, as the boats had raced past them. It had fired the imaginations of sporting men all around the world, that the stock and produce exchanges were interrupted with bulletins of the race's progress.

The boats were well matched; with eight boilers each and engines of comparable size, it was truly a race between captains. Thomas Paul Leathers captained the southern favorite, the 1547 ton Natchez. The Lee, slightly smaller at 1467 tons, captained by John W. Cannon, was the favorite of St. Louis, the East, and Europe.

Cannon and the Lee had won the race by refueling en route; the Natchez had lost time by having to stop for fuel. The race had been so fierce that Cannon was accused of shaving his head to streamline it!

A superb public relations effort, the pilot's report on the race established that a new record had been set for the distance: three days, eighteen hours and fourteen minutes; and that it took a day and thirty-eight minutes to get to Vicksburg.

Captain Leathers, the owner of the losing Natchez, rationalized that he had lost over seven hours of race time due to fog and engine maintenance, but then sportsmanship reigned and he joined the celebrations.

Bill and his friends joined the celebrations at the Southern Hotel. Alfred had served in the Union Army and had guarded the bier of Abraham Lincoln; he'd also been an officer in the Fourth Street Bank and now would help Bill with the bookkeeping and finance courses that were part of the curriculum at Jones Commercial College. As the three young men daydreamed of adventure over their beer, they had no way of knowing that their future in wiremaking would be an adventure in itself.

THREE

1871
Kansas

Sent by Frank Ludlow on a sales trip, Bill bought his Katy ticket at 2nd and Pine Streets to travel on the Missouri Pacific Railroad. The Katy had been designed to travel from St. Louis and Fort Riley to bring gold and silver from Mexico City. It was on this trip that Bill was to encounter Indians and the legendary Big John Scullin, the chief track layer for the Katy, and later the founder of the great Scullin Steel Company of St. Louis.

Bill had travelled to Parsons, Kansas, which was named after Levi Parsons, one of the founders of the Katy. It was in Parsons that Bill met the rollicking Irishman, Big John Scullin, who told Bill some tall tales:

"My men can lay track better drunk than many can sober," he'd said.

"The toughest part of my job is dragging my men out of the saloons to get them to work on time," he'd added.

"I'm bigger than most men, but there are some cobs I'd hate to tangle with. Let me tell you about my most frightening experience. I walked into a bar one night and one of my stronger boys was at the bar saying, 'I'll do anything I want in this bar. And if any one tries to stop me, I'll beat the shit out of 'em.' Well," Big John continued, his blue eyes sparkling. "I said, 'I'll stop you!' Bill, I was swinging but he was *hitting*. But finally I subdued him! Now we act like it never happened, but we give each other a lot of walking space!"

Big John also told Bill about the affects the Chicago Fire was having on the railroad, the Katy, in particular. "You see, most of the country's paperwork went up in that fire . . . a lot of the insurance money is going to fire damage, and the insurance companies are in trouble, even to bankruptcy. So insurance money that should be available to us is gone up in smoke!"

During this trip, Bill and Big John had a drink with Alfred Clifford who was on his way to catch a stage to San Francisco. "I'm going to try to drum up some wire and grill business," he explained to Bill. "From Frisco I'll be going by boat to Mexico and overland to Mexico City . . . a real whirlwind sales trip! I'll bring back a lot of sales orders and make a lot of contacts."

"Well," Big John said. "I wish you'd come up with a commercial barbed wire to keep those damned buffalo and longhorn off the tracks. Can't you do something about that?" He'd sipped at his whisky. "The prices are breaking us."

Bill had said, "I'm working on a hand cranked machine now. That will drop the price of barbed wire. Just be patient a bit longer."

While you're at it," Big John had said. "Why don't you make a barbed wire design that is individual to the railroads? That way with all our long, untended fences, when the wire is stolen, with the individual design . . . why, we could track the thief that much easier!" Then, ordering another drink, Big John had said, "Well, here's to you, boys. I'm laying my ribbon of rust along the length of the old Texas Trail . . . even with cholera and malaria, we're able to lay twelve miles of track in ten consecutive days!"

During this trip, too, Bill encountered Indians. The Kaw Indians were the only tribe in Kansas; their reservation consisted of 230,000 acres of land, their tribe numbered seven hundred people. The railroad ran through the Kaw territory, and one day at a place called Council Grove, the Kaws stopped the Katy that Bill was riding. Peaceably enough, they walked out onto the tracks and climbed over the engine, examining the ironwork, peering in at the passengers and the engineers. Bill, seated in the special investor's car, watched in astonishment as the Indians took the

1871 Wire Salesman in Kansas

flags that decorated the car. "What next?" he wondered. He'd heard tales of travellers being kidnapped and robbed and abandoned to wolves. He couldn't stop the shivers of dread that made him shake in his boots as he watched the mysterious and enigmatic Indians strip the flags from the train.

But then the diamond stack of the train sent up a shower of sparks, the train lurched into motion. The Indians chased it for a short distance, then, good naturedly, they waved it on. Bill sighed with relief.

Later, that land which had been taken by the U.S. government, was returned to the Indians by the courts and the Katy railroad lines had to give millions of developed acres to the Choctaws, Chickasaws, Creeks and Kaw Indians.

Bill spent the rest of that train ride trying to forget about Indians by planning new ways to improve wiremaking.

FOUR

1873
A Trip On The Natchez
To New Orleans

Bill worked very hard for the next three years. Then Oliver Saylor decided that Bill needed a vacation so he sent him for a long, leisurely trip down the Mississippi on the Natchez with Captain Thomas Paul Leathers.

Captain Leathers had built the Natchez in 1869; for part of every year, Captain Leathers and the Natchez hauled cotton to New Orleans for transshipment to overseas textile mills, the rest of the year he hauled passengers and freight. On this trip, which included some grill fence wire as cargo, Bill travelled along as companion to Captain Leathers in order to learn about the Mississippi and the people of the South who were still coping with the Reconstructionists.

Bill's stateroom measured twelve feet by fourteen feet, contained a bed, mirror, closet, wash basin and water pitcher. He would share the bathroom down the hall. He had brought his engineering books along from the Mercantile Library, and planned to study despite the constant shaking and chugging of the Natchez. At twenty-five, Bill was energetic, inquisitive, anxious to learn; he found plenty of space to exercise his long legs by walking the Cargo, Hurricane and Texas Decks; and Captain Leathers, always willing to talk, provided exercise for Bill's active imagination.

Captain Leathers told William: "The Mississippi River system is the longest in the world. It discharges twenty-five times as much water as the Rhine.

"It meanders so much that the river that LaSalle discovered has moved away to leave nothing but farmland and forests where it used to be rapid water!

"Once Ste. Genevieve was a port town! In the 1700's it was the port town for the mines of Phillipe Renault and de La Mothe . . . and now Ste. Genevieve is on a plain . . . separated from the river!

"Oh, and the early French were of noble descent . . . They could talk you into thinking their way about the craziest things . . ." Captain Leathers had grinned at Bill and said, "Why, they even tried to have their Catholic Diocese declare that duck is edible on Friday because it is waterfowl and waterfowl are a form of fish!"

Bill had noticed a very pretty young lady strolling about the decks of the Natchez. He introduced himself to her. She was Sarah Drain, the eldest daughter of John Drain, a prominent south St. Louis real estate dealer.

"I just inspected a piece of land myself," Bill told her. "I think your father's name was mentioned as the owner. I'm glad to meet you. Do you live in St. Louis?"

Eyes twinkling, Sarah told him that she was eighteen years old and that she was studying English literature at Washington University. "I was born in

Connecticut," she said. "But Papa moved here to St. Louis a few years ago. Papa had owned a windmill that powered the textile mill looms, and when we came here he got his start in real estate. Even in these troublesome financial times, Papa is doing well, I'm happy to say."

"I just acquired a lot on South Broadway," Bill had confided. "I got it for a thousand dollars from a friend of mine who sells for Papin Real Estate. I got the lot for that low price because the Credit Mobilier Group is siphoning funds from the Northern Pacific and that's making for distress in the business community." Bill had found himself talking animatedly to Sarah, and he thought, "She's a very attractive young lady." He'd said, "I guess you wouldn't be interested in such dry stuff as finances."

"Oh, but I would! My Papa talks about his business at home and we all are interested. And the papers are full of the bond forgeries and bogus shares everybody is being fooled with . . . why, those bogus shares are going wholesale! Even trusted names like the Erie and New York Central have had bogus shares sold!"

Bill nodded sagely, "Yes. Securities traders nowadays need to know their sources when they deal. You can't take anything on trust anymore It's a shame, the great expectations of the land grant Transcontinental Railroads look like they won't be coming about as hoped."

"You can't say that the financial world is dry *or* dull," Sarah had laughed. "Men always say that women gossip . . . but it's the men who make the scandals!"

Bill had laughed, too. "I know who you mean! Big Jim Fisk!"

"The Gorgons of Finance!" Sarah and Bill laughed together as Bill recited the newspapers accounts of the Big Jim Fisk scandals: Big Jim Fisk had had a mistress named Josie. And Josie had taken a lover, a social climber named Stokes. And Josie had boasted to Stokes about her love letters from Big Jim. Then young Stokes had attempted to blackmail Big Jim with the love letters. Big Jim had taken young Stokes to court. After the court trial, the enraged and disappointed Stokes, gun in hand, had followed Big Jim to the Grand Hotel. And there Stokes fired upon and mortally wounded the financial gorgon, Big Jim Fisk. It had been a delicious scandal.

Bill and Sarah looked up from their chatter as Captain Leathers came to stand near them. "I'd be pleased if you'd come to have lunch with me," he invited.

Arm in arm, Bill and Sarah joined the Captain and walked to the main cabin where lunch was served. Impressively furnished, the main cabin was more luxurious than many hotels. Imported from France, the furniture was heavy walnut, inlaid with a lighter wood, and handcarved, the initial "N" was carved within a holly wreath. The handpainted china had been designed especially for the Natchez; the silver engraved with a likeness of the Natchez on each plate. The Irish linen tablecloth and napkins bore handworked "N" for Natchez.

Sarah had looked around and said, "Everything is a feast for the eye." As indeed it was, for the skylight was of stained glass, and all the filigree work was of natural cherry wood.

The lunch itself was sumptuous: several courses were served, even two kinds of Sauerbraten. Indeed, Bill and Sarah admitted that they were stuffing themselves, eating up every morsel! "And, with these high prices," Bill said. "You have to eat everything you get so's you won't waste anything . . . money or food!"

The meal passed pleasantly, and Captain Leathers, always the story teller, pointed out a government snag boat. "Snag Boats," he said. "Have made the river a lot safer since the war. They are a marvel of modern engineering! Why, back in the heyday of riverboating the river was full of deadly embedded logs and roots that had come downriver. In fact, between St. Louis and here, there was a sunken hulk for every mile of river. There were two hundred of those sunken riverboats at one time. . .

"And it wasn't only the logs and roots that make the river dangerous. There are the shoals and the sandbars where a sternwheeler could get stuck if it couldn't push itself off. The Army Corps of Engineers planted bright lights on those shoals and sandbars to alert the sternwheelers at night. That was a help, too.

"See how the Snag Boat works? That winch on the boat lifts the log or whatever the obstruction is, out of the water. Then the log can be cut up and floated away. It was Henry Shreve back in the '30's that invented the Snag Boats. The government gave him a contract to make the Red River navigable. He did that and laid out Shrevetown, his camp for his workers. Before Shreve came along with these Snag Boats, the Red was jammed solid with logs.

"Right now, Shrevetown is struck with the yellow fever," Captain Leathers added. "You know, the Negroes in Louisiana swear that they can cure the yellow fever with their voodoo. And a lot of white folk believe that, too. Of course, once you get yellow fever, your brain is affected and you go fast."

Sarah changed the subject. "How do you know where the safe places are in the river?" she asked.

"My river pilot. He's the most valuable man aboard. And he costs the most! I pay him four hundred dollars a month. But he's worth every penny of it. He knows every kink in this river, and she's changing almost every minute! He gets all the latest reports on the changes, of the new snags, sandbars. He can tell by the swirls and the 'shinings' where the river bank is being eaten away by the river, or where a new sandbar is forming. He tries to memorize the river, his own special section of the river. He's got to be perfect in his knowledge or my boat will go down and be lost."

As they neared Memphis, Bill and Sarah asked the Captain about the Sultana disaster.

"That was a tragedy beyond telling," Captain Leathers said. He leaned back in his chair, his eyes thoughtful, sad, as he began to recall the events surrounding the Sultana's tragedy. "She'd come out of New Orleans on April 21 of that year, 1865. She was a beauty of a boat, big and new. She was rated to carry 376 people. But she only had aboard a hundred passengers and a crew of eighty or so. And some hogs, mules and about a hundred hogsheads of sugar, so she wasn't even up to capacity. Now this was about twelve days after Lee had surrendered, so things were calm then, no more shooting or explosions on the river or the banks. So they'd expected a calm trip all the way. But then in Vicksburg, the engineer reported a leak in one of the four boilers, so the captain ordered a layover for repairs.

"It was behind Vicksburg that the Confederates had rustled up a camp for delivering the Union prisoners. They brought them there, starved, just skeletons, from Cahaba and that hell-hole of an Andersonville. And as quickly as they could, the Confederates placed those prisoners aboard northbound steamboats. I guess they loaded about a thousand to a boat, overburdening and crowding, but you'd know these prisoners wouldn't care! And before the Sultana came along, the *Henry Ames* had onloaded thirteen hundred and the *Olive Branch* had taken seven hundred. There was talk that a Union officer was taking a bribe from one of the ship companies . . . well, no one knows . . . it was not proved. But to stop that kind of trouble, the officer in charge put the rest of the camp onto the *Sultana* even while two other steamboats owned by the . . . bribing company, were available.

"But those poor wretches from the camps came in wagons or on the little railroads, they came any way they could manage . . . on foot in groups!

"And at first the officers in charge tried to keep order, tried to put the men aboard in some organized way: the Michigan troops on the main deck, the Ohio troops on the hurricane deck . . . but the prisoners were so desperate, so hungry to get home, that the order crumbled and no one

The Sultana Disaster

knew where anyone was . . . and no one knew for sure how many were aboard! It was said that there were twenty-four hundred soldiers aboard as well as a hundred and eighty civilians.

"The *Sultana* made it to Memphis by around 7 p.m. and some of the troops helped to unload the sugar, while others, those that could, went over to the Whiskey Chute to have a drink. The lucky ones missed the *Sultana* when she got under way at midnight. She left for Cairo and was in the middle of those little islands, the Hen and Chickens when a boiler exploded." Captain Leathers hesitated for a moment, looking at Sarah. "It's a terrible story," he said.

But Sarah, caught by the proportions of the tragedy, nodded him to continue.

Captain Leathers cleared his throat, then went on. "The blast from the exploding boiler made the splinters and shards lethal, they knifed into the poor sleeping bodies on the main deck. The steam scalded God knows how many others. Sleeping men were thrown up into the air and came down in a rain of bodies and pieces of bodies." He shook his head. "Everybody that survived had a terrible story to tell.

"It's enough to say that no one knows for sure to this day how many were killed. One report says that over fifteen hundred were killed. The military says that *only* 1238 were killed. However many, it was too many.

"And because of Lincoln's assassination, and John Wilkes Booth's death, there was only a little insignificant newspaper story about it. Seems like they could've taken more notice than that of so many dying." Captain Leathers looked away from his guests. "I'd have said it was one of the great marine tragedies in history . . . or in times to come."

Captain Leathers was correct. For the *Sultana* tragedy claimed more lives than even the later *Titanic* tragedy.

When they reached Vicksburg, Sarah and Bill went sightseeing. They walked through the caves that had been used as shelters from the bombs Grant had fired upon the city. They visited the ice factory where water was placed in pans in thermally insulated rooms. Pipes of ammonia gas drew heat from the water, thereby cooling it. Bill explained to Sarah, "The gas is created by pumping liquid ammonia through a small opening, that expands it to a gaseous state."

Sarah smiled up into Bill's eyes. "It seems like magic to me! That you can form ice without winter!"

As they walked through the store rooms that were corklined to store the ice all summer, they licked the cold chunks of ice that they were given as a treat.

When they returned to the boat to continue their journey they met a garrulous old Missourian named Jim Putnam. He regaled them with stories about himself, how he ran a sawmill in "Pop" (Poplar Bluff) and had been born in "Jop" (Joplin). He come aboard to take a short vacation trip down the river. "I planned to visit Napoleon, Arkansas," he told them. "But the river washed the whole kit and kaboodle away! Schools, churches . . . even the bars! In only fifteen minutes that whole city that had browbeaten Greenville from across the river was washed away . . . only a single brick chimney left to mark the spot!"

Jim Putnam sat with Sarah and Bill during supper. He bowed when Sarah said goodnight to the gentlemen. Then he winked at Bill and the Captain and sang all the verses of "She Wiggled as She Wobbled," a song with many verses. "Fit only for men's ears," he chuckled.

He sat up all night planning his sightseeing in Natchez where he'd have a cool drink in the once brawling Mecca for flatboat drivers, the "Natchez-under-the-Hill". He told Bill stories about the flatboats, how they were brought downriver and sold as lumber; he told Bill about the hard drinking, hardy keelboatmen who drove the flatboats. "Gone now," he sighed. "Men of an earlier time."

Other colorful characters boarded the Natchez. Another of these was Dr. Sol Smith, who boarded at the Barrow plantation dock. "I'm out of the plantation business now," Dr. Smith informed Bill and Sarah. "Once upon a time I owned a plantation worth $200,000. Then the war came and the plantation lost value. Down to $10,000.00 and a $14,000.00 mortgage. No, I'm through with the plantation life. I'm off to New Orleans to practice medicine.

"You see," he added, "the only northern money being invested in the South now is in the riverboats and the railroads. The profits are quicker. So, the farmers, like me, couldn't afford equipment, no matter how large our plantations.

"Look over there," Dr. Smith pointed at a curve in the river. "That's called the Cote d'or, or the Gold Coast of sugar plantations . . . it brings in fabulous amounts of sugar and money." He pointed again, farther along the river, "There's the De La Barre plantation. It goes along the river for dozens of arpents . . . that's 192 feet to an arpent, little lady," Dr. Smith explained to Sarah. "And that plantation goes back from the river six miles clear to Lake Ponchartrain . . . it's 25,000 acres! A fantastic place! You should tour it sometime. That's a great old family. De La Barre crossed the Italian Alps centuries ago and settled the La Flelche area of France. When Francois De La Barre came to New Orleans in the early 1700's during the Spanish rule . . . he was made a member of the Cabildo that governed the territory. The family generally became notaries and lawyers, a high class family, as I say. But now since the war, I'm sad to say, like the rest of us, even the De La Barre's fortunes are in a decline. They're intermarried with the Barrow family of Tunica, Louisiana, but I don't know if that helps them much." He looked wisely at Sarah and Bill. "This is a good time to be young and starting out, like you young people."

Sarah and Bill blushed. They hadn't said anything to each other but it was obvious that a romance had kindled. Sarah had arranged to stay at the St. Charles Hotel so Bill escorted her there before returning to spend the night aboard the Natchez. Because of race riots, the streets were not considered safe, although the new governor, William Pitt Kellogg had declared the entire state under martial law.

While Sarah and Bill were in New Orleans they visited Wenger's on Bourbon Street and saw the "self-acting organ, the forerunning of the player piano". They were part of the overflow crowd at the St. Louis Cathedral where Marie Laveau was a daily communicant before she began the day as a voodoo priestess. They passed Congo Square hundreds of negroes danced modified French steps to African rhythms played on drums.

Bill, ever the engineer and planner, enjoyed the progress made on the repairs to wharves and streets directed by the Federal General Butler, declared by the southerners to be "the best scallywag we've ever had".

But he and Sarah were happy to return to St. Louis to watch the completion of Eads Bridge as it connected St. Louis with Illinoistown (now known as East St. Louis).

FIVE

1875
Barbed Wire Machine Designer

Bill and Sarah married in 1875 at the St. Patrick's Church. They took an apartment in the St. Claire Hotel at 3rd and Market Streets. The St. Clair had housed other illustrious tenants such as Abraham Lincoln, John Calhoun, Robert E. Lee, Daniel Webster and Charles Dickens.

The next few years were to be busy for Bill. Bill began to work on a hand cranked barbed wire machine but before he could patent his, his friend, Vaughan, had patented a similar design: the barb was wrapped around the fence wire but the barb point was sheared on by other means. Vaughan's machine was controlled by Elwood and the Eastern Trust of Washburn & Noen, one of the earliest wire manufacturers.

In order to compete with Elwood, other St. Louis manufacturers such as Ludlow-Saylor and Harrison Wire Company at 22nd and Papin Streets asked Bill to work with them as their designer. He was to produce machines to automatically wrap and point barbs. He worked at his designs during the second shift at the St. Louis Iron and Machine Works at Chouteau and Main Streets. At this time barbed wire was priced at 18¢ per pound and Bill believed that his machines would double the rate of production, thus lowering the costs of barbed wire. Working under contract, Bill designed a number of barbed wire machines to compete with the Elwood-Glidden type machine (which is still in use today).

When Bill had saved enough money, he travelled to England, France and Germany to increase his knowledge of wiremaking, even returning for a time to Altena where he studied with his old friends. Upon his return to St. Louis in 1877 Bill formed a partnership with Oliver Saylor, leased the works erected originally by Frank Ludlow, and installed additional machinery and improvements. Misfortune struck, and a disastrous fire destroyed the whole plant, taking all that Bill had invested.

Adversity seemed to spur Bill's inventiveness for next he turned to making wire loops for Ellis Wainwright's brewery. By this time the brewing industry was second only to textile manufacturing in the city. Bill and Sarah used the wire loops to fasten the corks on to beer bottles; they did this work at home in the evenings.

Bill sold the lot on South Broadway, using the $3000.00 from the sale as working capital to form a partnership with O.P. Saylor and a Mr. Wright, each of whom contributed $1000.00 to the enterprise. The agreement Bill made with Saylor and Wright was that Bill would act as plant manager for a trial period of one year, if, after that first year, Bill's management satisfied his partners, he would continue in that role, reimbursing the partners for any losses of that year. Of course, Bill was highly successful and his partners chose to continue the arrangement.

However, the wire works were on ground leased from the Iron Mountain Railroad who demanded possession. Since Bill had rebuilt the entire factory on the ruins of the burned down works, he now had to dismantle the plant and relocate it. Because his partners were not able to help him. Bill had to do most of this work himself. He did manage to secure a new location in fee and to erect a substantial new plant.

Not long after that, Saylor, who was in poor health, and Wright who was very old, sold Bill their joint interests for $11,500. The year was 1879 and Bill's new partner was Thomas W. Fitch, the president of the neighboring Harrison Wire Company, the largest in the city of St. Louis. Bill and Fitch each invested $6000.00; the company was incorporated as the St. Louis Wire Mill. The company grew and prospered, within two years it was the only plant capable of making the wire rods that were the raw materials for the wire drawing process.

Bill and Fitch respected each other as partners, still they had their differences. Fitch allowed Bill to use the Harrison Wire Mill after hours to work on the design of new machinery. Bill had worked in and around wire mills for sixteen years and had learned every detail in the manufacture of wire, so he watched every operation with an eye to cost and time efficiency. Fitch felt that since he had been in a supervisory capacity much longer than Bill, that questions about how a job was done were a waste of time; Fitch believed that completing a job on time was of prime importance and that cost analysis, time and motion studies and per unit cost analysis were the "wild dreams of theoreticians" and had no place in the practical world of manufacturing. Bill acknowledged that Fitch was an "older and wiser head" but he trusted his own experience and instincts and continued his rudimentary cost accounting, mostly after hours on his own time.

Fitch had decided that Bill was to manage the St. Louis Wire Mill. As Bill hired more people for the mill, he watched each operation and made time and motion studies of each operation. Bill was determined to standardize work procedures and whenever possible to implement job specialization rather than to continue the disorganized jack-of-all-trades methods that had previously existed. And when Bill found a man to be especially effective in his job, he made sure that man was kept happy salarywise so that he would not have to train a replacement.

Watching Bill supervise every detail of the wire works, Fitch told him, "Bill, I thought I was frugal 'til you came along. Sometimes I think your string saving habits are overdone, but your effectiveness is unbelievable. I have managed plants myself and have come to the conclusion that taking the time to figure per unit costs is a waste of effort; that even if you have this data, what can you do with it? But who can argue with success, Bill? You are a child prodigy if I every saw one; twenty-eight years old and the best manufacturing man in St. Louis. You've either got it or you don't."

While Fitch believed that Bill was the most efficient manufacturing man in St. Louis, he thought that Bill's analysis of the details was superfluous. Bill later proved Fitch wrong by becoming one of the first industrial engineers. And he could pay royalties to other wire manufacturers and still beat their prices in the marketplace. Bill was an admirer of another industrial engineer, the famous Frederick Taylor of Midvale Steel, who later wrote the book *Cheaper by the Dozen.*

In 1880, Isaac Elwood and the Eastern Combine brought suit in the court of Judge Blodgett to protect their 220 patents in the barbed wire field. Judge Blodgett found that their patents covered the utility of barbed wire usage. Blodgett's decision gave a virtual monopoly to Isaac Elwood and his friends. This judgement was to cause difficulty for the other manufacturers and to barbed wire users as well.

(No Model.)

W. EDENBORN.

MACHINE FOR POINTING WIRE.

No. 250,795. Patented Dec. 13, 1881.

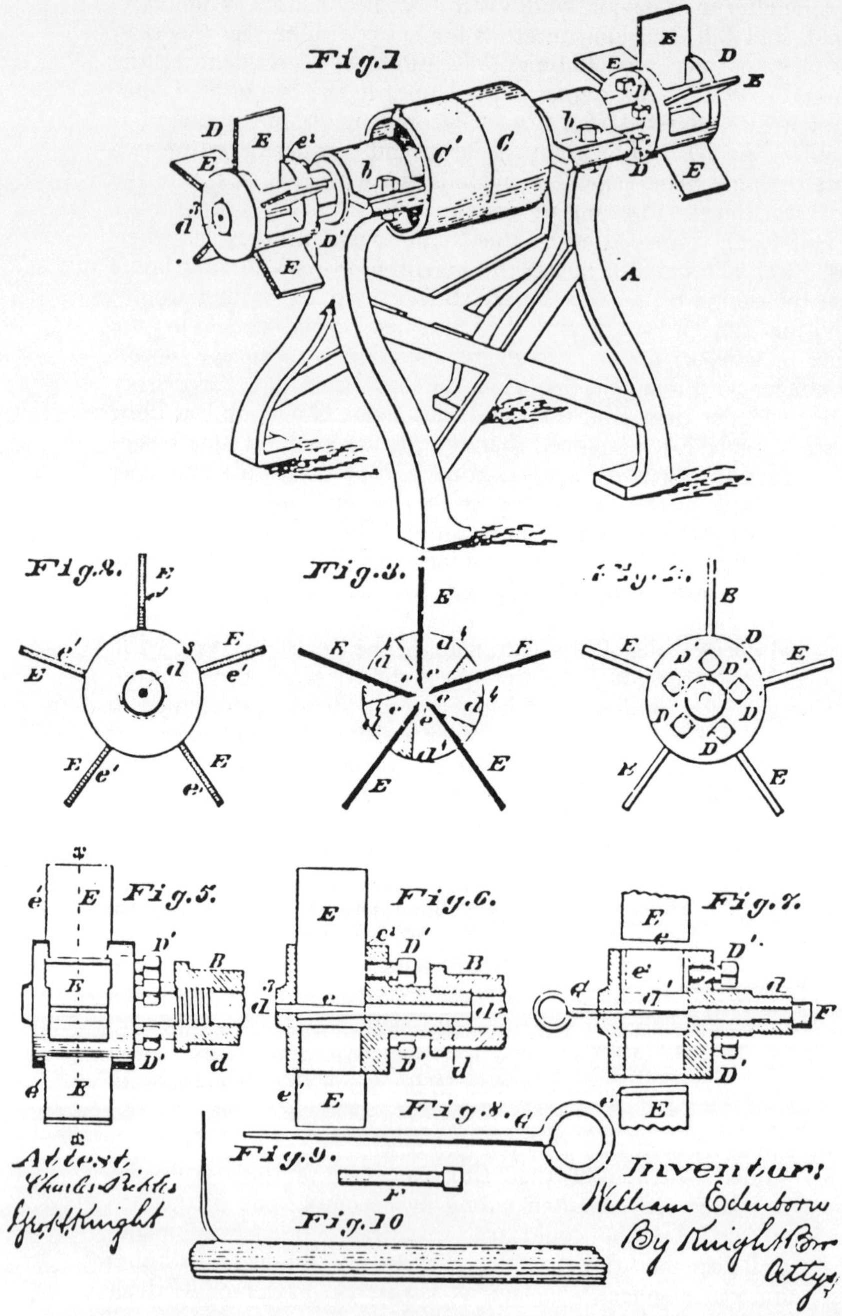

In December, 1881, Bill obtained his first major patent. This was the first of hundreds of laborsaving devices that Bill would invent. This was the rod pointing machine, variations of this machine would be installed in every wire, wire rod and wire nail mill in the country. The operation is simple with 5 rotating knife blades cutting a sharp point on a wire, nail or rod. The machine is switched off by moving the rotating belt from driving, wheel C to unloaded wheel C. Pin G is used to calibrate the knife blades. This is patent 250,975 and can be obtained from the Patent Office for 50¢. Prior to Bill's invention, the part to be pointed had to be rotated, requiring more exacting adjustments.

One evening during this time, Alfred Clifford, and his wife, Mary Francis dined with Bill and Sarah at Tony Faust's. When Bill asked Clifford how his sideline of barbed wire was doing, Clifford told him a story about his new partner and full time manager of the small shop on Laclede's Landing, John W. Gates. Gates had been hired originally by Isaac Elwood, the barbed wire monopolist; Gates had demonsrated the efficiency of barbed wire to control longhorn cattle for ranchers in San Antonio. Then Gates had left the Elwood Barbed Wire Company to work for Clifford's moonlight barbed wire factory. Clifford was a fulltime officer of the Ludlow-Saylor Wire Company, and was aware of Gates' sub rosa activities which were the infringement upon Isaac's many patents. Isaac had sent the Missouri process servers to serve an injunction against Gates; However, Gates had learned of the action and had loaded his equipment on a barge and floated it to Illinoistown. When Isaac sent the Illinois officials after Gates, he had taken to the barges again.

Bill, hearing this story, knew that Isaac had the financial power to make it very difficult for any new entries into the barbed wire business. He decided he'd design his own barbs and automatic machines and avoid the wrath of Old Isaac. Sarah agreed, saying that they had enough trouble on their hands just settling their own household.

The brewing industry was booming by now and Bill and Sarah entertained friends often at Schnaider's Beer Garden, two blocks south of Lafayette Park. Schnaider in 1852, had opened a brewery there which in its first year had produced 3000 barrels. Three years later the establishment had been sold to a competitor, Urban & Hammer, who borrowed $9000.00 from Eberhard Anheuser, a soap manufacturer from Brunswick, Prussia. Anheuser then foreclosed on what was at that time called the Bavarian Brewery. At the instigation of his friend and a brewer of Humboldt, Kansas, H.C. Brendecke, also from Brunswick, Anheuser bought out the other creditors and became a brewer. Later in 1861, Anheuser hired Adolphus Busch as Sales Manager and the young Busch married Anheuser's daughter, Lilly. In 1875 the two formed the Anheuser-Busch Brewing Association, with a capitalization of $240,000. They bought the picture "Custer's Last Stand" and the first copy went to Custer's Regiment's Chow Hall in Fort Riley, Kansas. Bill had encountered Indians himself just south of Fort Riley at Council Grove.

The use of barbed wire to control cattle did not come about easily. In the middle 70's when Gates and the DeKalb entrepreneurs Isaac Elwood, Jacob Haish, and Joseph Glidden began to sell barbed wire many farmers felt it was a "rope from hell" and would be cruel to their animals. Then range wars erupted when the cattle bosses realized that the small farmers or "nesters" could use the barbed wire to enforce their boundaries and to encroach upon the open grasslands. Gunmen were hired, men like Billy the Kid appeared on the ranges. Fence cutting was made a felony in some areas. And some fence cutters were found wrapped in barbed wire.

The non-vicious type of wire was in great demand during the middle 1880's and Bill's web-type wire had a good market.

Eventually, however, the range cattle learned to respect the barbed wire and to live with it. But before that happened, a catastrophe occured in the Texas Panhandle that brought a public outcry against the "inhumane" wire. The northern part of Texas had been fenced from east to west in taut 4 and 4 wire sections measuring 10 to 40 miles in length. This had been done to keep strays from wandering south and devouring the precious grass. In 1885 there were tens of thousands of cattle north of this fence when a snowstorm hit. The cattle instinctively headed south but couldn't get past the fence. They piled up against the fence but their huddled bodies couldn't make enough warmth. Frozen, they piled against the fence for as far back as 400 feet. The fence had done its job too well.

Eventually, however, barbed wire was accepted even by the large cattle barons. Henry Sanborn, one of Bill's manufacturer's representatives, along with the inventor of many basic barbs, Joseph Glidden, fenced and cross fenced 250,000 acres of Texas Panhandle to enclose the first large cattle ranch of the west. Sanborn also laid out the city of Amarillo, Texas.

One of Bill's customers was the Capitol Syndicate of Texas who had fenced the greatest ranch of all time: the XIT (Ten Counties in Texas). The Syndicate had built the Texas Capitol Building, which was to be one foot taller than the national capitol, in exchange for three million acres of high plains Texas Panhandle stretching from an altitude of 2000 feet to the altitude of 4000 feet. It took seven years to fence the "Ten in Texas". The ranches had special wire so that any pieces stolen could be identified; the railroads who had long untended fencerows also had special wire that could be identified by one square strand.

By now the factory had grown to a valuation of $103,000.00. They made a capitalization of $100,000.00 and Mr. Fitch, although entitled to one half, insisted that Bill receive $51,000.00, or a majority of the stock.

They had also made a trade alliance with the St. Louis Wire Fence Company. Bill had acquired an interest in the Western Union Wire Company, of which he was elected president; Mr. Charles Hale was the secretary and treasurer; Oscar Moore was a stockholder. This enterprise, under Bill's management, also prospered.

Bill competed with men like John Gates and Isaac Elwood. While they did not have the technical knowledge that he had, they were aggressive promoters who were willing to take chances on something that still had an uncertain future: wire. He felt overwhelmed by these men since most of his time was spent researching technical information and developing new ways to make wire and they, a freewheeling breed, seemed about to dominate the industry. But he remained confident about his own abilities and his ability to establish a future for his own family.

Sarah and Bill now had a daughter, two year old Antoinette. They planned to teach her the ABC's early because they believed there were no "smart" people but only people who spent more time and effort training their minds. Bill remembered his own hours of pouring over engineering and technical information at the Mercantile Library and some of the educated and intelligent people he'd met there: Adolphus Busch and James Eades.

This was a happy time for Bill. He had his family and his own business. He was in control of his own destiny. He enjoyed looking at the majestic Eads Bridge which symbolized what technology could accomplish.

Bill was hard at work designing another barbed wire machine. But this time, he thought, it's as complicated as McCormick's reaper. Bill thought about Edison and Edison's search through 6000 materials and all the trial and error before he found that tungsten would burn for more than two days without disintegrating. "But Edison had lab assistants to do the materials tests," Bill muttered. Bill had to do all the work on this barbed wire machine himself. Except for help from his assistant Kennard, Bill had to test every rotating hammer, every lever, arm and pawl himself. Secrecy had to be maintained because the competition between barbed wire manufacturers had become cut throat and spies were a very real threat to anyone who dared to design a machine.

While Bill worked in the evenings on his new machine he thought about how the new machine could benefit people who were trying to fence their lands at a cheaper price; often the cost of the barbed wire fencing was higher than the land it was supposed to fence! And other worries concerned Bill: he desperately wanted security for Sarah and Antoinette. With the constant pressure of competition in the wire business, prices might fall, his competitors might capture some of his business; and while the wire rod mill was producing most of the sales, the raw material from Europe was exorbitantly high priced.

And even as he worked at the factory on the new machine he thought about what his attorneys had said about the machine: that it was a "departure from prior art" and if he could construct a working model they would get him letters patents which could lead to royalties and the much sought after financial security. The Knight Brothers, his attorneys, assured him that once they had the working model they'd know the right people to approach, they'd do the politicking for him.

So those evenings at the factory, when Sarah came down to keep him company and grade her English papers, Bill worried about the possibilities of a depression, or another panic like the Railroad Panic of 1873. And he brooded about the possibilities of Carnegie trying to steamroll the wire market.

But it wasn't all worry for Bill. Saturday evenings he spent at home with Sarah and Antoinette. And as Sarah played the piano, Bill would recall the classic definition of engineering: the art of directing the great sources of power in nature for the use and convenience of man. And he knew that his final contentment would be in his contributions to society. He thought about Goethe's frantic Faust who had sought contentment in every worldly pleasure only to find near fulfillment in a project to reclaim land from the pestilential swamps. And as Sarah played the piano those evenings, Bill lost himself in the contemplation of how he would eventually fulfill his visions.

The notorious barbed wire decision of Judge Blodgett in 1880 had given the Ellwood-Washburn combine a monopoly in the barbed wire fence field. Now, in 1882 another decision was made that changed the situation dramatically. Judge Treat of St. Louis (one of the founders of Washington University) declared that the claims of the Ellwood-Washburn combine were too vague and too broad. The Treat decision gave the St. Louis manufacturers the right to continue with their manufacture of barbed wire. The proximity of St. Louis to the expanding Southwest gave the city an advantage over the more distant competition. The Treat decision prompted Bill to hire the noted machine expert, Gustave Griesche, to assist him in his efforts to design a two strand wire with barbs on each strand, and to design another machine to make wire with barbs on one strand.

That year, the Harrison Wire Company, one of the casualties of the Blodgett decision that had given the Ellwood-Washburn combine the wire monoply, was dissolved. The president, Mr. Fitch, retired and sold his interest in the company to Bill.

This was a time when Bill was able to watch business change hands, indeed, he was able to form a partnership with some of his own best friends. One evening while Bill and Sarah dined with Alfred Clifford and his wife, Clifford told Bill the story about how he got involved with Ludlow-Saylor, one of Bill's biggest competitors. A Mr. Dean, a large investor in the Ludlow-Saylor Company as well as the Harrison Wire Company, had gotten tired of all the bickering at Ludlow-Saylor and wanted Alfred Clifford to buy him out for $10,000. Alfred had given Dean $1000 of his savings and had borrowed the rest to buy into the company; he believed the stock could be sold later for $30,000. Alfred said that he could put up with all the policy altercations and bickering at Ludlow-Saylor, but he would sell if he found a suitable buyer. Bill said that he wanted Clifford and J.W. Gates to join him at his wiremaking company instead.

When the Clifford and Gates Southern Wire Co. Plant at 805 North Second Street burned down. Bill invited Clifford and Gates to join with him. They arranged the business this way: Bill would remain in the background as plant manager and President. Bill and Clifford would handle the finances and establish credit to buy and construct new plants; Bill had the production and borrowing power and the others had the knowledge of sales, finance, and accounting to supplement him. Gates would be the sales manager. Bill had a large royalty income and had the borrowing power for expansion, and this organization would allow Bill to concentrate on plant operation and production. Bill would retain 49% of the stock, Clifford 26% and John Gates 25%.

The main plant occupied the block bounded by 21st, 22nd, Papin and Gratiot Street (presently behind Union Station). He acquired more land at Braddock, Pennsylvania (near Pittsburgh) and laid out a wire rod plant which became the most efficient in the country. Bill was fortunate to have built this plant during the depression of 1884 when construction costs were down, thus this plant provided a source of less expensive wire rods.

Bill's engineering has contributed greatly to dropping barbed wire prices from 18¢ per pound to 3¢ per pound. The barbed wire output of the St. Louis Wire Mill was all that Ludlow-Saylor and Shapleigh & Simmons Hardware Companies could handle.

Bill continued to work in the plant and to design new wiremaking machinery. As much as his men admired his abilities, Bill admired Frederick Taylor even more, and often discussed Taylor's cost controls and analyses with his friend Alfred Clifford who was part owner of the Ludlow-Saylor company. Bill continued to invent barbed wire machines, believing that if the Treat decision was overruled, he could still sell his designs to the combine.

The demand for the "vicious" large barb was diminishing in the face of the demand for the short barb which was easier on the cattle. The "Steel Trust" of Isaac Elwood and Worcester firm of Washburn and Moen had control of the basic Glidden patent, the short barb. The Glidden fence had a barb with a second stranded wire; Isaac and his friends had control of most of the other barbed wire patents and most of the machine patents. Until this time they'd done most of the barbed wire business in the country; in 1874 the industry had produced and sold 10,000 pounds of barb, by 1880 that had increased to 80,000,000 pounds. The Steel Trust and Isaac Elwood were having difficulty policing their patents since J.W. Gates had succeeded in violating them with his "moonlighting" across the Mississippi. Not only were there a lot of barbed wire moonlighters in the industry but now Bill had contributed

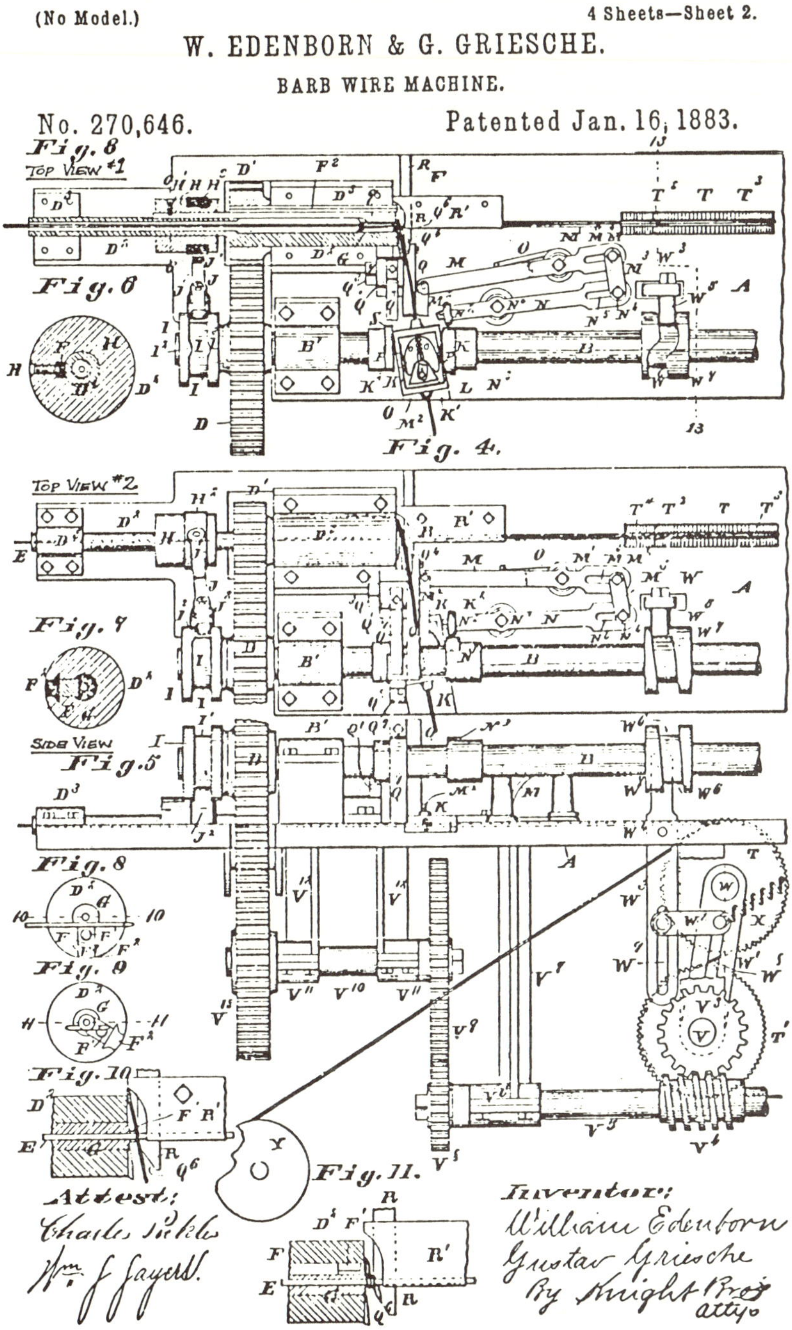

Operation of Patent

The operation of the "machine that fenced the West" is simple. The main fence wire is pulled from Reel T. This wire is halted every 18 inches, or so, by a synchronizing mechanism, while the transverse, or barbed wire, is wrapped around the fence wire. The transverse wire is pulled in at the proper time by metal fingers at K. The barb is sheared on at Q, and the wire is wrapped by hard metal at R. This machine can make the Glidden (popular) wire at a rate of more than one barb per second. Barbed wire is still made in this manner. An automatic machine, such as W. Edenborn's, made possible the fencing of the West. This barbed wire is stranded with a barb wire in a subsequent operation.

many competing machines. Many of the barbs didn't require or need a patent because the machines that made them were so unique that they were easily identifiable, and now many of the barbed wire fences could be made on contract. And now, due to the Treat decision, the St. Louis barbed wire makers such as Bill and Gus, would be able to take over the barbed wire business from the Steel Trust because the St. Louis factories were closer to the markets of the southwest. Also, the Confederate veterans who settled in the southwest were more easily able to identify with the St. Louis people, rather than northerners from Worcester, Mass.

After much puttering, trial and error, Gus and Bill developed patents for two machines to make the Glidden wire, both machines were patented January 16, 1883. Patent No. 270,646 produced barbs on one strand of wire, another strand was added later. This machine, known as the machine that fenced the West, was their producer and enabled the St. Louis Wire Mill to announce a price cut that shook the competition. Patent No. 270645 was a more complicated machine but produced a barb on each strand.

By this time Bill had his own barbed wire, No. 271,693. The unique feature of this fence was that the barbs grasped both wires tightly so that if one wire broke between the barbs, it continued to be held securely by the remaining barbs so that the rest of the strand was unimpaired. The automatic machine needed no patent. Patent No. 299,763 on a barb for wire fences required a very simple machine to fabricate. Bill's barbs were short, not hurtful to the cattle. Patent No. 313,929, now a collector's item, has a large cross section visible to cattle, it consists of two wires with a web between. There was no welding or soldering, the web was obtained by rolling a large wire lengthwise. Also, it was believed that a single fence wire would be less expensive than a two strand wire, so Bill invented a machine that would hammer a kink into a strand, with the kink holding the barb in place.

These inventions saved the consumer fortunes and made Bill very wealthy; Sarah and Antoinette enjoyed their frequent trips to the factory, even with wealth, they continued to come to work with Bill and to live frugally near the mill itself.

Bill, Alfred and his friends were able to buy out their less efficient competitors because Bill was able to produce barbed wire at 3¢ per pound. Bill purchased control of the Lambert & Bishop Wire Fence Company and met John Lambert, a one time warden at Joliet Prison. While the warden, Lambert had put down a prison riot by locking the prisoners in their cells, closing all windows and doors, turning on all the ovens and furnaces which caused such heat and stench that future riots were forestalled. Bill also bought the Iowas Barb Wire Company of Allentown, Pennsylvania; he bought a St. Louis competitor; Bill bought the Banker Wire Company of Lockport, Illinois; he also acquired an injunction bound plant of the DesMoines inventor, George Baker. Bill installed his own machines which were outside the control of Isaac Ellwood's patents. And, Bill was able to pay for most of these acquisitions in cash!

Alfred Clifford had discovered while studying their raw materials that although prices for billets were low in England, there was a shortage of raw materials in America. It was decided by them that John W. Gates, the master negotiater of the company, would go to J.P. Morgan, who was attaining prominence as a banker, and borrow money, based on the St. Louis Wire Mill credit, and buy most of the available stocks of raw material. Gates was able to persuade Morgan, he bought the raw materials from England and sold it at a profit to the American mills with plenty to

(No Model.)

W. EDENBORN.

MACHINE FOR POINTING WIRE.

No. 250,795. Patented Dec. 13, 1881.

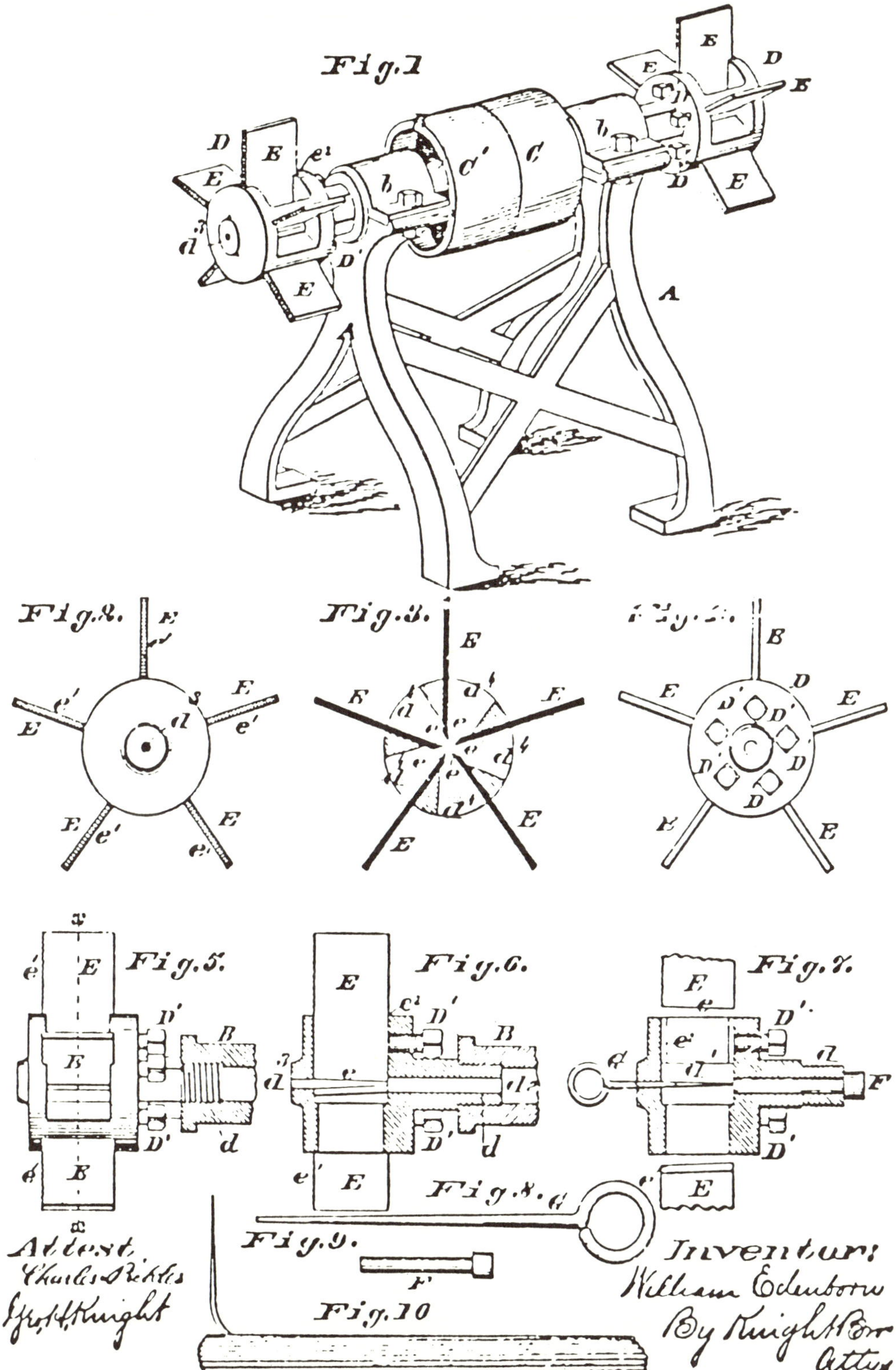

Attest,
Charles Pickles
Geo. H. Knight

Inventor:
William Edenborn
By Knight Bro
Attys

spare for the St. Louis Wire Mill. This adventure was considered a bold coup by the business community and made Gates and his backers famous.

Bill had another project that would bring him further success. His rod pointing machine was incorporated into a cold-forming process for making wire nails. The nail would be cut, headed, and pointed automatically. The cheapest way to make nails until this time, was by shearing them out of metal and they were square shaped with no head. Bill's wire nail machine would drop the price of nails from 8½¢ per pound to 2½¢ per pound and would revolutionize the industry. For seventeen years during the life of his patent, Bill received a royalty on every wire nail produced. Revolutionizing the construction industry, William Edenborn's inventions would eventually lower the national price levels considerably. His visions of helping mankind were becoming real.

For relaxation Bill would journey to the Dardenne Duck Club across the Missouri River where he was often the guest of Edward Mallinckrodt, the Ammonia King. They would travel by stagecoach along St. Charles Rock Road and cross the river by ferry.

Sarah and Antoinette often accompanied them and stayed overnight in St. Charles, Missouri. Antoinette was the pride and joy of Bill and Sarah, and was considered to be beautiful and precocious by the teachers at the Mary Institute of St. Louis. Because Bill was so often away on business these week-end trips were a special family time for all of them. A fond father and devoted husband, Bill told Sarah and Antoinette that they were his only friends. He and Sarah planned for his semi-retirement from business so that they could spend more time together.

One of the few intimates to this family circle was Aunt Nettie Drain who was a student at Harris Teachers College. She wanted to teach near her home in South St. Louis where her parents had built a Victorian home on 3649 Cleveland Avenue, the first home west of Grand Avenue. Nettie often entertained Bill's family by reciting Robert Burns poems. Bill often quoted the lines, "The best laid plans of mice and men aft gang agley," especially when he'd contemplate the way his first wire mill had burned down.

SIX

1890
Park Avenue

In 1890 Bill moved the family to their new home on 2019A Park Avenue, facing Lafayette Park. They had a new neighbor, Dr. Emil Preetorius, an 1848 Revolutionaire and the founder of the conservative German newspaper, THE WESTERN POST. One of the best educated men in the area, Preetorius hired men of equally high calibre, such as Carl Schurz and the young Joseph Pulitzer who had come looking for a job. Preetorius hired both men to work as reporters on his paper: Schurz had already established himself as a prime mover of the German people and had been a General in the Union Army during the Civil War; Pulitzer, young and brash and confident had wanted "any job, I'm not particular".

Bill and Sarah enjoyed their family as often as Bill's busy schedule allowed. Antoinette, now ten years old, precocious, was always at the head of her class at the Mary Institute. And they had adopted a one year old, Lilly, to be their baby daughter.

Antoinette read everything she could get her hands on and had read Thomas Hardy's *Tess of the D'Urbervilles.* Bill had read it, too, and was grateful that his daughter, Antoinette, was off to a better start than the tragic Tess.

In addition to her studies, Antoinette enjoyed travelling with her Aunt Nettie Drain. They travelled to Europe on the majestic *Frederick the Great*, took driving trips to Scotland and Germany, with special visits to Altena and Westphalia. When they travelled in the United States they toured the Mississippi River and the southern states. Bill and Sarah would accompany their girls to Europe whenever possible, usually when Bill had business concerns to occupy him, such as the royalty payments from his wiremaking machinery.

Then, when Antoinette was fifteen, tragedy struck. Bill wrote this letter to Gustave Griesche:

"My daughter Nettie, whom you knew as a child, by the time she was fifteen years and three months of age, met with an accident and lost her life. I was proud of the girl, as she had been nothing but the first in her class and was, at the time of her death, attending the Mary Institute Branch of Washington University. She had grown up rapidly. I encouraged her taking outdoor exercise after school hours and she was quite proud of her riding horse. The old cable road covering the street passengers by way of Park Avenue. And my girl was on horseback in the eastbound track and noticing that approach of the car behind her, urged the horse to leave the track. But as a whole block of houses had been in the course of construction with piles of mortar, brick, rock and piling placed along the avenue, the horse, though gentle, did not want to leave the car track, in fact walked sideways, front feet out of the track and rear feet between the rails. The motorman simply cranked the bell, though noticing that the girl was trying to get the horse out of the track. She noticed the stubbornness of the horse not wanting to move, and by her urging him, she finally threw him around trying to escape the other way. But by that time the car was upon her, and the momentum may be imagined, when the impact killed both the girl and the horse. This was January 5, 1895."

Bill and Sarah had intended to retire to their farm in Chesterfield, Missouri, but instead buried Antoinette there. They took Lilly and moved to Chicago to live at 5524 Washington Street. They could not stand the sound of streetcars.

Edenborn Home - 2019 Park Avenue

SEVEN

1891
The Largest
Wire Company

By this time Bill was constantly involved in battles over patents in the barbed wire industry. When, with his attorneys, Bill won a multimillion dollar settlement from the Steel Trust, he decided to buy the rest of the patents in the fence industry. With Gates and Clifford Elwood, Bill formed the Columbia Patent Company. Because the patents of 1874 and 1875 were running out, Bill knew that this was a good time to bargain with Isaac Elwood and the Steel Trust.

However, the Columbia Patent Company was doomed to failure for several reasons: first, there were too many patent infringements from outsiders and Bill could not control them; second, there were too many diverse interests at conflict within the company itself; in addition the press attacked the Columbia patent arrangements: THE CHICAGO TRIBUNE wrote, "The barbed wire men have finally formed a trust.

They do not call the combination a trust, however . . . it was decided after considerable discussion to speak of the organization as a pool."

Although the Columbia Patent Company did fail, Bill learned from the failure and formed the Consolidated Steel and Wire Company, of which he was the principal owner and President. Bill had learned from his attempt to pool with Elwood and the Steel Trust that future pools must be strictly policed and that heavy fines for forfeitures must be exacted from firms caught cheating within the pool. This knowledge made Bill much more cautious and scrupulous in his dealings with other firms. And, although the Columbia Patent Company had failed, Bill had attained control of most of the wire nail and fence patents. Now, the establishment of the Consolidated Steel and Wire Company was heralded by IRON AGE MAGAZINE as, "The most important single event in the wire industry.

"There is no doubt that the Consolidated Steel and Wire is the largest maker of barb wire and wire nails in the world."*

Bill's vision for the company was at once grand in scope and yet complete down to the tiniest detail. First, the company planned to free itself of dependence on suppliers of raw materials; it would acquire its own mines, blast furnaces and railroads. Bill and Alfred Clifford planned to stay in St. Louis. Gates and Lambert would move to the Rookery in Chicago; the Rookery, which housed many of Bill's executives, was a marvel of architecture by Burnham and Root, an eleven story structure, East Indian in design, with marble corridors, onyx walls and an impressive spiral staircase.

For his legal counsel, Bill hired Elbert Gary, son of a Methodist minister. Elbert H. Gary was destined to learn the steel business and to make his way to the top of the industry. Bill said to Gary, when he accepted the retainer to become the general

*The real beginning of U.S. Steel.

counsel to the company: "Judge Gary, you are new to the steel and wire business and are as good a corporation lawyer as there is in the country. We need you here in the Rookery with us on a retainer basis as we anticipate further expansion and we need someone with your perspective not to mention your relatives in the Eastern Establishment. You seem to be a person with no faults who will go far in this business."

My friends and I in the shop are working on machines to make a new type of woven wire fence which will eliminate the hurtful barbs. With the royalty income I hope to buy out more competitors. These larger combinations seem to be the trend in spite of the Sherman law. A larger concern will cut down destructive competition, enable us to decrease the cross shipment of goods, add more research and development. We can now afford our own overseas sales office at 18 Billeter Street, London.

'A larger concern is manageable as evidenced by the success of Frick and Schwab of Carnegie Steel. We hope you stay with us, Mr. Gary."

The courts now decided that a combination that affects trade does not necessarily restrain it. This type of adjudication resulted in hundreds of agglomerations of companies. The largest of these agglomerations was the Morgan, or money trust, followed by Standard Oil Trust, which consisted of J. Rockefeller, E. H. Harriman, and Bill's friend, Henry Rogers.

These lean years from 1893 to 1898 were good opportunities to buy out competitors and to expand with relatively inexpensive labor. Bill realized that now, more than ever, his own companies' independence required that they have their own mines and raw resources. And, within ten years of the Judge Treat decision, Bill, Alfred and their friends controlled the fencing of the west! With their expanded tack production at the Braddock and Joliet Mills, and continuing to buy new plants, Consolidated Steel and Wire was making 75% of the annual sales of barbed wire products in the United States.

And even with the enormous successes of his enterprise, Bill still had time to plan for his workers. One of the first things he did as president and controlling shareholder of the Consolidated Steel and Wire Company was to establish a company-paid retirement and accident plan for the workers. Although he met with opposition from the top management, Bill insisted that this plan be administered. Later, other manufacturers followed Bill's lead and established their own workmen's plans.

Once his daughter had said to him: "Dr. Preetorius says you are one of the richest men in the country, Daddy."

And Bill had responded, "Even so, money wasn't meant to be spent in a spend thrift fashion on ourselves, Nettie."

Bill did not fail his own vision of helping others to have a happy and safe life.

But then, tragedy struck again. Their youngest daughter, Lilly, was stricken with diphtheria; and although everything was done to save her, their Lilly died. Bill was forty-six years old, Sarah, thirty-eight; it was doubtful that they would have any more children. They clung to each other in their personal lives and became partners in their business life so that they were rarely apart. And, if anything, they became a bit more reclusive.

And yet, these were to be the golden years of expansion for Bill and his friends in the wire industry. Bill had solved the problems of lost motion and inefficiency in coiling wire as it came out of the wire dies. His wire coiling machines and variations of them are still in use today. Some of the patents still in use are: 429,511; 429,512; 480,565 and 480,564. He

(No Model.) 4 Sheets—Sheet 1.

W. EDENBORN.

MACHINE FOR MAKING WIRE FENCING.

No. 558,787. Patented Apr. 21, 1896.

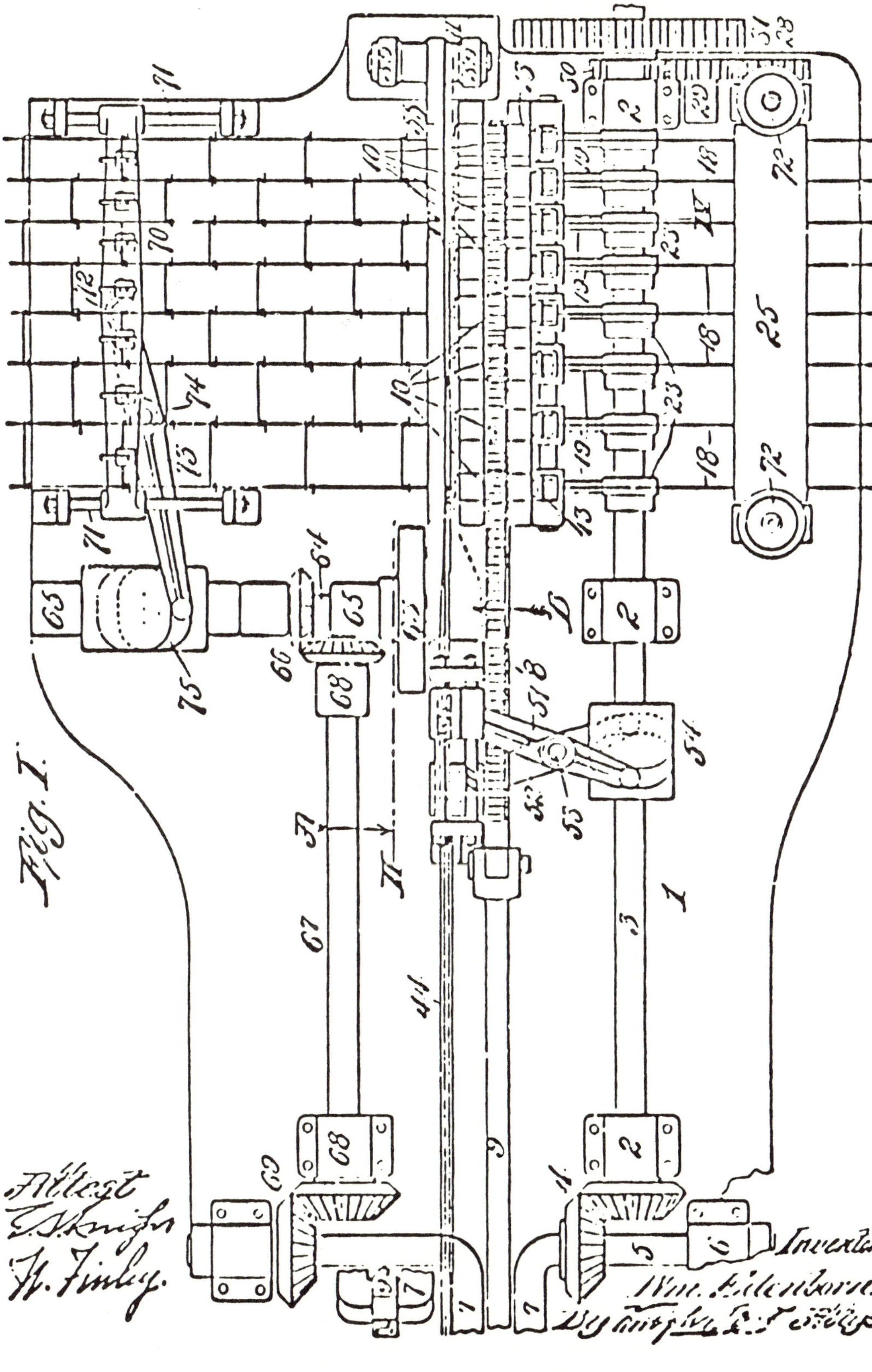

Fig. I.

Attest
E. S. Knight
H. Finley.

Inventor.
Wm. Edenborn.
By Knight Bros. Attys

also designed a cheaper and lighter reel to hold wire and wire fences #543,801. Bill had seen the early wire factories and their lack of organization, that lack had prompted Bill's insistence on job specialization and job organization; he'd seen too many men lose an arm or leg to the uncontrolled wires that whipped around on factory floors. It was this scrupulous eye to detail, added to his inventive genius and his drive, that brought William Edenborn to the Gay Nineties, a man about to see many of his visions come true, and to see millions of dollars earned. Bill's machines to make hog fence* and woven wire fence gave the Consolidated Wire Company complete dominance in the wire fence industry.

Bill and Judge Gary worked closely together at this time to expand Consolidated Wire's holdings. Although they differed on some subjects (Bill believed that the workers should be rewarded by bonuses, Gary believed that the workers were adequately rewarded by regular pay) they agreed that Consolidated should expand only under stable conditions. And they were both appalled by Carnegie's aggressive price cutting. And Gary learned from Bill how to close down old, inefficient plants while retaining existing management.

Business was good for the Consolidated Wire Company and for Carnegie Steel. This was subject to much discussion by Bill and Alfred because their competitors were sufferng through a panic that had begun in 1893. Prices had declined steadily since the Civil War. Bill realized that this decline was due to an increase in the supply of goods, that increase made possible by engineers and inventors such as himself. But the other and more destructive reason was that the government had begun to redeem greenback dollars with gold, decreasing the money supply.

Then, when McKinley was elected President, Bill and his friends decided to speed up their acquisition of options for competing wire mills. In addition, they would acquire rod mills, because the rods were the raw materials they needed.

It was during this time that Bill's friend and associate, J.W. Gates acquired his nickname of "Bet a Million." An insatiable gambler, Gates would bet on anything. Bill recalled that once Gates and Isaac Elwood had made bets on how many flies would land on a cube of sugar. They'd had pencils and scorepads and the bets had gone up to $1000 per fly! Then, Bill recalled, there had been the time aboard a train when Isaac and Gates couldn't find a deck of cards so they had bet $1000 on drops of water as the drops moved down the window pane to the bottom. "Bet a Million" got his nickname from losing a million dollars on the drawing of a single card. As Bill said, "He was rambunctious even as a child!"

With his own daughters gone, Bill became fond of Bet-a-Million's son, Charlie, and employed him as his secretary. While "Bet-a-Million" was tough with competitors, he had spoiled his son, and now Bill had his attention occupied by the rambunctious son. Charlie entertained himself at the Everleigh Sisters' famous bordello by drinking champagne with the fancier girls. The Everleigh girls dressed well, behaved politely, and tried to educate themselves by learning to read. According to Charlie they were educating the wrong end. But Bill saw that Charlie was not "too dumb to get himself into trouble" so he tried to keep Charlie out of mischief by dictating business letters to him.

Now Bill had begun another venture. He had started to build a railroad down the Red River Valley from Shreveport. Bill had written to Clarence Ellerbe the head of the Louisiana Central Construction Company because the lumber resources of that area were still largely undeveloped.

*Note the raw wire going in on the right and the hog fence output on the left.

Bill thought that the long and short leaf pine would rebuild the South, and that the longer growing season in the South would make growing the seedlings easier. Bill had already begun to work with landmen, surveyors and other agents to buy the rights of way and whatever cypress lakes (brakes) were available. Cypress was especially useful for railroad ties in the river wetlands because it was impermeable to rot. Bill's imagination, always active, had already decided that he could build and equip the railroad for $40,000 per mile. With Ellerbe's assurance that he would have the right of expropriation (Eminent Domain), Bill hoped to gain the right of way for his railroad at a fair price. In addition, Ellerbe had told Bill that he would receive tax subsidies from the parishes and that the large landowners would give him grants. Another way of financing his railroad, Bill believed, would be the sale of lumber F.O.B. Shreveport; and he would plat out a town every few miles and sell the lots to raise even more money for his new venture. He had even laid out a railroad town and named it after Arnsberg in the Ruhr Valley where his parents had taken him as a child. This Arnsberg, in Louisianna, would be a town devoted to rest and relaxation, a vacation town for the railroad workers and the people of Shreveport. Ever shrewd, Bill had even given town lots to the local politicians to enlist their support.

EIGHT

The American Steel & Wire Company and J.P. Morgan

By 1898 Bill and his friends had acquired options to most of the competing wire companies. E.H. Gary had done most of the legal work to establish what was to be the first American Steel and Wire Company. Most of the financing had been arranged by Henry Seligman and Joseph Strauss, with the famous corporate attorney, Max Pam doing additional legal work. J.P. Morgan had planned to finance some of the acquisitions, but with the sinking of the battleship *Maine* and the subsequent drop in the stock market, Morgan bowed out of the project.

In 1899 the American Steel and Wire Company of New Jersey was formed. Bill was the principal owner and the president of this concern which had an output of over 2,500,000 tons per annum, greater than even the output of Carnegie Steel. The executive committee consisted of Bill, John Gates and Isaac Elwood.

Gary, Bill's former associate, had now convinced J.P. Morgan into financing Federal Steel and absorbing Illinois Steel. This combination then emerged as the second largest steel company, following closely after Carnegie Steel. Gary became the chairman of Federal Steel and coordinated the mines, railroads, ships and plants that came under his domain.

It was during this time that J.W. Gates undertook one of the first of his many sub rosa moneymaking schemes. Gates had been a member of the executive committee of the original American Steel and Wire Company, and had taken advantage of his inside information to form a stock tracing pool which had brought him a sizable but not quite legal fortune. Had Bill Edenborn known of this he would have been furious at such a betrayal. But Gates was not caught at that time. Later, he would be caught at something else.

And even now, while organizing such a monolithic industrial concern as the American Steel and Wire Company, Bill still had insisted that workmen's compensation plans and Old Age Plans be installed. Once while he was appraising a mill before he acquired it, he was approached by a worker who said, "Mr. Edenborn, we hope you buy this plant. Our present retirement plan is 'crawl in a corner and choke'."

Bill believed that his retirement plan and compensation plans benefitted not only the worker but the company itself. Example: Billy Garret who had invented Consolidated's first continuous rod rolling mill at the Joliet, Illinois plant. The operation involved rolling square billets down to wire rods by means of a roller train. Billy had gotten too close and one of the white-hot-serpentine strands seared off one of his legs. Bill and Sarah had brought Billy his claim check, visited him in the hospital, and when Billy recovered, he went back to work improving his machines. So, Bill reasoned, he had saved a good mind for the company as well as saving a man's life.

THE IRON AGE listed Bill's holdings in the September 7, 1899 issue:

"Official Statement Submitted to Stock Exchange by THE AMERICAN STEEL & WIRE COMPANY

At a meeting of the committee on Stock List of the New York Stock Exchange, held last week (about Sept. 1, 1899), the $40,000,000 preferred and $50,000,000 common stock of the American Steel & Wire Company were admitted to the regular list. The official statement submitted by the officers of the company contains paragraphs of the organization, which are of interest at this time.

Comment is made upon the fact that the company own the following plants:

PLANTS ACQUIRED, OWNED, AND CONTROLLED

American Steel & Wire Company, Illinois, plants at Anderson, Ind.; De Kalb, Ill., 2; Evanston, Ill.; Joliet, Ill., 2; St. Louis, Mo.; Cleveland, Ohio, 2; Findlay, Ohio; Allentown, Pa.; Beaver Falls, Pa; Rankin, Pa.

Washburn & Moen Mfg. Company, plants at Worcester, Mass., 2; Waukegan, Ill.; San Francisco, Cal.

Worcester Wire Company, plant at Worcester, Mass.

Cleveland Rolling Mill Company, plants at Cleveland, Ohio; Newburgh, Ohio, and iron mine property at Negaunee, Mich.

Indiana Wire Fence Company, plant at Crawfordsville, Ind.

Garden City Wire & Spring Company, plant at Chicago, Ill.

Consolidated Barb Wire Company, plants at Joliet, Ill., Lawrence, Kan.

Laidlaw Bale Tie Company, plant at Joliet, Ill.

Cincinnati Barb Wire Fence Company, plant at Cincinnati, Ohio.

Union Rolling Mill Company (only), plant at Cleveland, Ohio, known as Emma Furnace.

Portage Iron Company (part) plant at Duncansville, Pa.

Newburgh Wire & Nail Company, plant at Newburgh, N.Y.

Allegheny Furnace Company, plant at Allegheny, Pa.

Pittsburgh Wire Company, plant at Braddock, Pa.

Shenango Valley Steel Company(part), plant at New Castle, Pa.

Oliver Wire Company, plant at Pittsburgh, Pa.

Oliver & Snyder Steel Company, plant at Pittsburgh, Pa.

Shoenberger Steel Company, plant at Pittsburgh, Pa.; also coal land and coke ovens in Fayette and Westmoreland counties.

Puget Sound Wire Nail & Steel Company, plant at Everett, Wash.

Edgar Zinc Company, plants at St. Louis, Mo., Cherryvale, Kan.

Puritan Coke Company, plant and land at Baggaley, Pa.

Puritan Store Company, store at Baggaley, Pa.

Clark & Sauntry, mine at Virginia, Minn.

Altena Mine, mine at Virginia, Minn.

Cuff Iron Company, mine at Iron Mountain, Mich.

The plants are all owned in fee except the following:

Shoenberger Steel Company, American Supply Company,
Edgar Zinc Company, American Mining Company,
American Coke Company."

Bill became involved with oil almost by accident. It happened this way: Arthur Stillwell, a St. Louis insurance man, sent orders for steel and wire to Bill from time to time. Once he joined Bill and Sarah for dinner in Chicago and told them about his schemes for a railroad. He had surveyed the map of Texas and had found an undeveloped area; he'd secured investors, mostly Dutch, and had decided to dredge the lake, Sabine, low, like the Zuider Zee, down to the sea in a large swampy area there. He'd formed a development company to build a city there, named after himself, Port Arthur, Texas. He had secured the right to buy the right of way to commence a railroad from Kansas City, Missouri 778 miles due south to Port Arthur. But for some reason, his project was losing money and now he wanted Bill to invest in the railroad.

While Bill and his friends decided what to do about the almost defunct project, Arthur Stillwell had gone to E.H. Harriman who owned control of the Northern Pacific. E.H. decided he would take an interest in Stillwell's railroad and bring it to completion in Port Arthur, Texas.

But Bill, relieved of this project, looked at his own railroad which he'd been building for the last four years and was confident about its future.

Because Bill had planned to transport steel products on his own railroad, he was interested in the completion of Port Arthur. Then, because evidence of surface petroleum had appeared at Beaumont, Texas, on the Kansas City and Southern Railroad, on a hill like structure called Spindletop Hill, Bill and his friends formed a company to take a chance on the area. Spindletop became responsible for 15% of the world's oil production. And the Texas Company of Port Arthur later became the Great Texaco Oil Company.

But now problems arose for Bill. He and his friends had installed J.W. Gates as chairman of the American Steel and Wire Company in January, 1900. Gates owned several million dollars worth of wire stock which made it reasonable that he hold a position so responsible. But orders had declined and it was decided that Gates would issue the instructions that closed one fourth of the plants while the other plants continued to fill purchase orders. This instruction from Gates put 10,000 men out of work. But Gates' real mistake was selling wire company stock short before closing the plants and allowing the newspapers to find out about it. That news caused a depression in stock prices and Gates became the central figure in the public scandal.

The financial managers, Seligman and Strauss, came to Bill's office to request Gates' resignation. A special meeting of the Board of Directors was called and Seligman and Strauss offered their own resignations if Gates did not resign. Rumors were rampant: one of the Directors who might have voted against Gates did not appear at the meeting; it was said that Gates had given that Director shares of New Amsterdam Power and Light. Gates finally submitted his resignation.

Alfred Clifford was elected the new Chairman of the Board. Not long after his election, Clifford came to Bill to tell him that J.P. Morgan had made a formal offer to buy the wire company at book value to be incorporated into an expansion of the Federal Steel Company. Clifford had refused the offer.

But that was not the last they were to hear from Mr. Morgan. Several months later, after rumors that Morgan had purchased the giant Carnegie Steel Co., he requested another meeting with Bill and his wire friends. At this meeting Morgan said, "Gentlemen, I have purchased Mr. Carnegie's Steel Co. and I control much of your source of supply. I will

Founder of Texaco

Alfred Clifford
Director of
Kansas City Southern

give you twenty-four hours to accept our offer at book value. If you don't accept my offer, I will form a competing wire company and see to it that you go broke."

Bill said, "Our company is similar in size to Carnegie's, what did you pay him?"

"That is confidential, Mr. Edenborn."

Gates, who was still with Bill and his company, wanted to know where he'd fit in if they sold.

Morgan replied, "Your financial antics preclude your taking a part in Big Steel."

"Mr. Morgan, I do the same things you do, only you do it behind closed doors."

"That, Mr. Gates, is what doors are for."

When Morgan left them to decide, Bill asked, "Does he really mean it? He controls the source of billets we make the wire rods from."

Chairman of U.S. Steel

Alfred Clifford said, "He has a lot to say about our shipping rates."

They decided to sell at book value of $100,000,000. Carnegie, whose concern was slightly more integrated, received almost five times book value or $492,000,000.

Bill received $50,000,000 for his share of the company, and after selling his patent rights to U.S. Steel, the largest trust ever formed, he had a net worth of approximately $90,000,000.

Gary had pushed Morgan into his largest financial effort and as chairman, Gary wanted his former teachers, Bill and Alfred Clifford to be directors of the board, and to serve on the executive committee and advisory board. But Clifford's eyesight was failing and he chose to remain only as a director. Bill didn't mind that he had probably been outbluffed, he knew that he had been out-bargained. He was satisfied that his wire company produced 25% of Big Steel's sales, and did 90% of the wire related activity such as nails, tacks and staples, and that his wire company did 100% of Big Steel's barbed wire, hog fence and woven wire fence business.

But they only received 7% of the $1.4 billion purchase price!

Carnegie gave a lot of his money away, endowing libraries around the country. Bill decided that since he was no longer in the wire business, he would develop the resources of Louisiana and develop jobs for the people in that area.

Bill and Sarah moved to 29 State Street in New York City; they had homes in Biswell Springs, Arkansas and in Alexandria, Louisiana. Bill continued the direction of the Louisiana Railroad, but his major investments remained with U. S. Steel where he was placed on the Executive Committee and the Advisory Board.

Bill became friends with the first president of U. S. Steel, Charles Schwab. Schwab had come up from the Bessemer furnaces at Carnegie Steel. Bill and Sarah dined often with Schwab and his wife, Rana, at Gentleman Jim Corbett's on Herald Square.

As president, Schwab had many differences of opinion with Gary, Bill's old friend and associate who now worked for J.P. Morgan.

Gary was the chairman of the executive committee, and as such, was as powerful as the President, Charles Schwab. Charles Schwab was accustomed to running things himself; Gary believed in committees where no individual had much power. Schwab was a leader of men, accustomed to negotiating with labor; Gary believed in a uniform labor policy for all the companies that had come under the control of U. S. Steel.

Eventually, Schwab bought his own small steel company, Bethlehem, for $7,500,000. He also bought the I-beam which U. S. Steel had rejected. Grey, the inventor of the I-beam, had worked at Cleveland Rolling Mills, a subsidiary of the wire company. He had submitted the I-beam to U. S. Steel, but the finance committee, headed by Alfred Clifford, had rejected it, believing that the special plant needed to cast a unit beam would be too expensive. U. S. Steel continued to fabricate I-beams by riveting three pieces together. Schwab and Bethlehem bought the I-beam and sold it at great profit.

During this time, Charlie Schwab was building a castle on Riverside Drive; it would be patterned after the castle, Chononceau, of the Loire River Valley in France. It would cost $7,000,000 to build, have 90 bedrooms, a 60-foot swimming pool, and an 8-lane bowling alley in the basement. With four elevators and a self-contained coal powerplant, it would require twenty servants to care for it. Bill and Sarah said they couldn't wait for an invitation to see this castle.

It was during this time that Bill, Charlie and A. Clifford saw the "expert juggler" from London, England who had changed his name from W. Claude Dukenfeld to W.C. Fields. If he made a mistake with his fast moving "objects d'art" he blamed that on an "old judo injury which recurred while ascending Kilimanjaro." He claimed this with a voice that "even as a child had a certain raspiness."

From 1901 to 1904 Bill was on the Executive Committee and advisory board of U. S. Steel; from 1901 to 1909 he was a director of that same huge company. His company was the world's largest with 213 steel mills and transportation companies, 78 blast furnaces, 41 iron mines, a fleet of 112 ore barges, 57,000 acres of coal and coke properties. And Bill's wire company produced 25% of its sales.

The directors and owners ran the entire country. Frick owned most of the coke which powered the blast furnaces; John D. Rockefeller owned large parts of the Mesabi Iron Range; Morgan controlled the finances and other industries, especially the railroads; Bill knew the wire business. Others in Big Steel were Oliver of Pittsburgh; Phipps of Pittsburgh; and Henry Rogers who managed Rockefeller's Oil Companies.

These men thought so much of Bill that they named an ore carrying supership after him; that ship serves as a breakwater for Gordon Park, Cleveland, Ohio.

It was during this time that "Bet a Million" Gates was up to a lot of shenanigans meant to harass J.P. Morgan. First, he had purchased a controlling share in the Tennessee Coal & Iron Company, the second largest competitor to Bill and U. S. Steel into buying him out for premium prices. However, Gates' timing was bad for this was during the Knickerbocker Bank Depression of 1907 and Edenborn and Gary went to Teddy Roosevelt, managing to convince Roosevelt that only a merger of the two large companies would forestall a national panic. With Roosevelt's agreement that the consolidation would not violate the anti-trust laws, Bill and Gary bought Gates out, but not at the premium Gates had intended.

Next, Gates, as frontman for an investor group attempted to get controlling shares of Colorado Fuel and Iron, but Rockefeller, who controlled Colorado Fuel and Iron managed to avoid the takeover. Gates had thought by controlling Colorado Fuel and Iron, that he could coerce Morgan into buying, again at a premium.

Gates did manage to get control of the Louisville and Nashville Railroad, and Morgan did have to buy that from him at a premium because the L. & N. posed too much of a competitive threat to his other railroad, the Southern Railway. Bill Edenborn had some shares in this railroad, too, so he had a vested interest in Morgan's welfare.

Gates who was not always scrupulous in his business dealings had met his match in the J.P. Morgan who was widely known as a financial gorgon. Morgan, just as dynamic a businessman as Gates, and who also did business "behind closed doors" had taken a liking to the honest, frugal Bill Edenborn. When Gary and Morgan laid out the model steel town, Gary, Indiana, in 1906, they asked the modest Bill Edenborn to speak at the dedication, a mark of highest respect.

NINE

1898
The Beginning of the "One Man Railroad"

Still grieving the loss of their children, Bill decided to take a less active role in the wire business, become a director of the company, and devote more time to developing the lumber and agricultural resources of Louisiana.

Bill's railroad, the Shreveport and Red River Valley Railroad, was one of the first in the South to haul logs by rail. Although there was a demand for logs in the South, Bill faced difficulties with railroad promoters in the South, who wanted to block Bill's enterprises and his railroad's entry into the south.

However, always resourceful, Bill decided that as he was able to bring finances from the wire company to invest, he would purchase Cypress Lakes (Brakes) which weren't of much value to the local farmers. Then he would use the cypress for railroad ties, because the cypress was rot resistant and would serve exceptionally well in the wet, swampy areas. Bill also planned to build lumber mills, and to use the fresh running water of that area to power steam engines as well as to float logs.

When Clarence Ellerbe, Bill's manager, reported that Bill's projected path through Grant's Parish was blocked by the promoter, Gould, Bill decided to bridge the river at Grand Echore north of Natchitoches, Louisiana. Bill was to find many obstacles such as the one posed by Gould, in his endeavors to build railroads that would haul lumber and business throughout the South. But he believed that with the subsidies offered to his railroads by the large landowners and the taxes from the parishes, that his railroads would eventually not only stop losing money, but would grow up with the country and benefit the people the railroad served.

To supervise his railroads more closely, Bill and Sarah built a home at the St. Maurice plantation, near the general offices at Shreveport. And because they enjoyed the free passes Bill's rank as president of the railroad provided, Bill and Sarah entrained for a long trip to a new wire plant (Puget Sound Wire) at Everett, Washington. They found the city to be a garden spot; they enjoyed the main street, Yesler, which had once been a chute for the giant logs; they rode the ferry to the Olympic peninsula; they listened to victrolas play Hawaiian songs like "Aloha Oe"; and they watched the operation of other railroads.

Bill realized that his closest competitor, the succesful lumberman, William Buchanan, might be able to limit Bill's own development of the area in Louisiana. Bill knew that if Buchanan's railroad went south from Stamps, Arkansas, west of Bill's railroad lines, toward Alexandria, that Buchanan's railroad would hem Bill in from the east. And that the Texas Pacific Railroad, controlled by Gould, would hem Bill in from the west. Bill understood that if he allowed his own raillines to be hemmed in by the other railroads, that his would be limited to the role of feeder lines, and never attain the profits or effectiveness that Bill had envisioned.

Bill had a vision for his railroads: that they bring not only lumber and produce to the people of Louisiana and Texas, but that they bring work and opportunities to the impoverished of all social classes, the working man as well as the gentry. In order to see his vision fulfilled, Bill had to buy land as cheaply as possible, to take a number of risks with his own hard earned money, and to compete with already established and powerful railroad interests. The following letters show Bill as he worked to compete with such powers as the Gould railroads, the Missouri Pacific (Iron Mountain) and the Texas Pacific; in addition, Bill's fiercest competitor was the St. Louis and Southwestern Railroad (Cotton Belt) managed by John Edson.

It is interesting to remember that Bill Edenborn, who led a quiet

life, never made the headlines of newspapers, who was frugal, modest, and unlike Bet-a-Million Gates, avoided scandal and notoriety, dared to tackle the railroad giants, and successfully competed with them.

During the next few years when Bill was entrenched in battling the larger competing railroads, he wrote letters to his associates that clearly delineate his battles and reveal the tenacity of his vision for the Louisiana and Texas lands and people. The letters are written from the wire company offices at 975 The Rookery, Chicago, the St. Louis office at 21st and Papin Streets, the Drain home at 3649 Cleveland, St. Louis and various hotels in the United States and Europe.

This first letter concerns two railroad promoters, Bernheimer, a Shreveport grocer and St. Louis businessman and his partner, Parker, who wanted to build a railroad from Greenville, Mississippi to Texas. During this time the best location to build a bridge over the Red River was at the bluffs of Grand Echore north of Campti, Louisiana.

"Shreveport and Red River Valley Railroad Co.
January, 1898

Gent:

I had Mr. Bernheimer in today, he who is to help Parker find money to build the R.R. from Monroe to Natchitoches. The scheme is for Bernheimer to get J.P. Morgan (to) back him in building from (the) end of Southern R.R. Greensville on the Mississippi to Monroe to Natchitoches and Texas-ward. He will go to New York at once but I am convinced he will not get Morgan to back him; and he says he will not put in any of his funds. There are plenty such. I put cold water on Winn Parish and Natchitoches Tax Subsidies for his Railroad, but told him to get backing but to include my road as I was anxious to sell and go to Europe. I think October I will come and find no construction for Bridge and Grandechore, thereafter?

William Edenborn"

In the following letter Bill's efforts to buy land and to keep the land prices down are shown. He also attempted to cross Gould controlled Texas & Pacific tracks in Shreveport. Later he bought a plantation from Cochran, his landsman, and named the plantation, "Luella" after Cochran's daughter.

"The Rookery
Chicago
1/31/98
Dear Clarence:

Replying to yours of the 26th and 27th, duly received, will say that it seems the St. Maurice plantation cannot be secured very promptly; probably it can be had later on, but the swamp land which Mr. Cochran was to secure an option on for you is also kept very quiet, which is something I cannot understand. Of course, I realize that you have to depend upon Mr. Cochran, the landman in the matter, and I trust his assurances to you will be carried out in good faith. Glad to note that Major Dunn has assurances as to right of way from Red River Parish line to Campti. I should judge Coushatta downward, where land ought to be very cheap, we should succeed in getting free right of way. If we do not I would say that we complete our road to Coushatta and then ostentatiously talk of relinquishing our efforts to build further, and if they see that we do nothing for a few months the others may reconsider. Possibly we might even have to do something toward surveying a line from Coushatta to Winnfield via Lake Village.

I return tracing and mark on same with blue pencil place where I should judge we could make connection with Texas and Pacific Railroad. It is of course a matter for the engineers and T. & P. to agree to. We should have a connection with them by all means. The T. & P. have taken a sensible view in the matter. They are aware of the fact that we have a legal right to cross their tracks. They are aware that some of their tracks in Shreveport on Commerce St. have but a doubtful existence; probably some of them have no authority to be there at all. They do know that the City Council has granted us the right to lay our track along the street, so they make a virtue out of a necessity. If they were to raise many objections they might endanger some of their tracks on Commerce St. without hope of defeating us in the end."

Bill complained about exorbitant tolls to cross the Red River Bridge from Bossier City to Shreveport. The bridge was owned by a Mr. Harvey who lived in New Orleans.

> "Regarding bridge toll demanded by Mr. Harvey, it does seem as if $2.50 a car was an unreasonable figure, yet as you say we cannot help ourselves just now. We had however better try to get whatever cotton seed we bring from our line only to the oil works in Bossier so as not to give Mr. Harvey anything on the cotton seed. Probably you might see the people in the Bossier oil mill and have them go down to the planters and contract for such seed as has not yet been shipped."

Throughout his many battles and attempts to outwit his competitors, Bill had to stress secrecy about his own plans when he wrote or communicated with his associates:

> "I note what you say regarding gravel. I wish you would not send any engineer down there to make any investigations, in fact in such cases where on your visits to any one place you find special land, agricultural or timber or any other valuable strata, such as clay, gravel, sand, don't pay any specific attention to it. I am afraid your talk to Cochran has been repeated by him to others, probably unwittingly but it may defeat our securing the tie timber in the swamp lands and probably also the St. Maurice farm. When I get down there next time will soon discover what there is to be had. When next in Shreveport it seems I will have to go from there to New Orleans to settle the matter with Mr. Harvey."

Bill, always the manager, made recommendations about the problems of bridge building at Loggy Bayou, south of Shreveport:

> "Consolidated Steel & Wire Co.
> January 29th, 1898
> Dear Clarence:
>
> When I was in our Shreveport Office I have heard it mentioned that sunken logs would be encountered. No one knows this to a certainty but it is talked of so much that everybody seems to believe it as an existing fact. Personally, I believe that if we crossed Lake Bistineau we might find sunken logs in same, but in Loggy Bayou, which seems a narrow channel through which the lake empties itself, I do not believe that sunken logs will be encountered.
>
> I hope the temporary bridge structure will, as you so think, be erected at a nominal cost and serve our purposes and be much help to us in getting our material to the right places. I fully realize that the moving of broken stone, sand, cement and bridge materials is of much moment, and that a temporary bridge might save us considerably. By the time you have all the bids in, you may prepare for the letting of contracts. When all are ready, I should be pleased to have same mailed to me first, though probably I may be down there and take the matter up with you, I yet hope that the track will down to Loggy Bayou by February 10th or 15th."

In order to build the railroad, Bill had to do some negotiating with the parishes where his tracks would be laid:

> "Regarding Winn Parish tax. I note you received my telegram which you seem to think may do very good as it will assure the St. Maurice people. You seem to think that if we build to St. Maurice and thence down the river we would get some saw mills to build up through the timber. Probably we could agree to build to St. Maurice, say within 14 months, and agree to build to Winnfield within two years and six months, or probably three years, in which case we might get some arrangement with some saw mill plant located at or near St. Maurice, and we run for them a line up to Winnfield. Thereby we would give satisfaction to both St. Maurice and Winnfield people, though I will repeat that the twenty odd miles from St. Maurice to Winnfield would do us much more good if in the direction of Alexandria."

The next two letters show Bill as he sold some investments to raise money for his new railroad lines, and as he attempted to keep the operating expenses down:

> "I leave tonight for Chicago and have an option out on my interest in the Cliffs. Expect to be called to Kansas City if anything comes of the option, though I am not sure that it will."
>
> "In the employment of executive heads of Departments in the Consolidated Steel & Wire Co. I have always followed the principle to ascertain the compensation an applicant expected, and if in need of such services to engage him and then determine as to his proper compensation by the results achieved. The road I am constructing is short, a little above 40 miles in length, and necessarily is not big enough to pay large salaries. The construction is progressing fairly well, considering the rainy spell they have had the last six months, and will be completed about June 1st. Everything going into the construction of the road is of a solid, substantial nature, and I shall expect a minimum operating expenses on the earnings the road may have.
>
> Yours truly
>
> William Edenborn"

Bill had to write letters of encouragement to his associates, to cheer them on:

> "The Rookery - Chicago
> February 2, 1898
> Mess. McIlvried and Ellerbe
>
> Gents:
>
> Now Boys, I am trying to get into a financially strong position and I am working hard to that end! You must not loose heart if things don't always come our way. Some day they will. Don't worry about high water coming. Go right on and push your grading and do as if everything down to Alexandria had to be done within 18 months and as if the Winnfield branch was only a side issue. The harder we push, the sooner our competition will lay down.
>
> Yours in haste,
>
> William Edenborn"

Bill received this letter from Clarence Ellerbe about a man who wanted his son to enter the lucrative nail business:

> "Louisiana Central Construction Co. Ltd.
> General Office
> Shreveport, La.
> March 8th, 1898
>
> Mr. Wm Edenborn
>
> Dear Sir:
>
> -:-Nails for Mexico-:-
>
> Capt. Pete Youree of Shreveport has asked me to get the following information for him regarding nails going into Mexico.
>
> The freight rate from New York and St. Louis on nails and wire to the principal Mexican points and the rate from the same points to the same points on nail wire. The duty on wire imported into Mexico and the duty on nails.
>
> He explains the reason he wished this information is that his son, young William Youree, expects to put up a wire nail mill in Mexico and to get some kind of concession from the Government in regard to lowering the duty on wire very materially and raising it on nails. It appears that this young gentleman has just returned from a trip into that country.

Mr. Youree would like to know from you, when you come down, the cost of nail machinery, operating, etc.

If you have no objection to my giving this information will you kindly write New York and St. Louis for freight rate or advise me and I will do so.

I thought it best to write you before giving Captain Youree any infomation; expect the market price on wire which he could have obtained from anyone.

Yours respectfully,

Clarence Ellerbe"

Bill always had his eye on every detail of his businesses. In the following letter, he makes changes that will save money for the company:

"Dear Clarence:

Mr. Rose had informed me that the Illinois Central Railroad would take our flat cars from East St. Louis to South Chicago to be loaded with steel rails for our Road at Bossier City, thus getting the cars down free of charge to us. Found to my astonishment that Mr. Rose had ordered the cars shipped via the Iron Mountain and probably it will cost us freight to get them. Found that 13 flat cars were not yet shipped and told Missouri Car & Foundry Co. to forward same to Illinois Steel Company, South Chicago, to be loaded with rails for you, and thus to get free transportation for the cars. This I trust will meet with your approval. At any rate, this was my understanding and I have had no information changing the matter.

I expect to be in and near St. Louis until probably Saturday night or Sunday, when I shall likely return to Chicago.

Yours truly,

William Edenborn"

Bill continued to add miles to his railroad tracks and to make suggestions for townsites and sawmills:

Consolidated Steel and Wire Co.
May 14, 1898

Gents:

Prudhomme lands with Springs. I call your attention to my former statement that at that point where Railroads for Saw Mills would leave Main road northwardly and probably another road would diverge toward Winnfield; at such junction there would be a really good chance for a townsite and running Spring of good water, and with a Saw Mill nearby would have fine chance of success. Of course, if this was known it would spoil the whole thing; Campti would not like it, etc."

And of methods to compete with the land promoters Case, Fordyce, and Beardsley, Bill wrote:

"As you know Case and Fordyce are *in* Beardsley's road; during lengthy talk a map was wanted and Welsh pulled out Louisiana pocket map to show country between Baton Rouge and Simmesport. I saw on the map lines from Bienville to Alexandria and to Trinity! If I could build a cheap line from Grappes Bluff to Winnfield, I would probably stop them.

"I favor purchase of Rubieaux Lake at $2000 if you think that is cheap. I hope you will get low option on gravel lands. Had we not better incorporate a Land and Lumber company say nominate Ed Drain President to ask prices on lands? Will consider further about election in 3 and 4 Natchitoches. Invite again Monday 16th.

Yours,

William Edenborn"

Costs were always important:

"Dear Clarence:

I am still waiting for your estimate for the completion of the road to Coushatta at least; of course if you can add thereto the constructing down to Campti would like to have it."

The relative costs of flat cars:

"The Rookery

Dear Clarence:

I enclose herewith carbon copies showing that I have paid the Pullman Palace Car Company for three coaches, and have ordered same forwarded to Bossier City, all in accordance with your desires in the matter.

I have also made payment to the Missouri Car & Foundry Co. of St. Louis, for their first shipment of 15 flat cars, $69000.00. I am in receipt of a bill for ten box cars which no doubt I will have to pay for at an early date."

The following letter describes in detail what Bill wanted his stations to be. The flag stations mentioned were not regular stops, the train stopped there only if flagged down.

"The Rookery
3/21/98

Dear Clarence:

I take it that no station is placed at the Loggy Bayou bridge, yet it seems to me there ought to be a station at that place. Possibly Des Arc or East Point could be moved a little. We should have a water tank likely at Loggy Bayou bridge so that the one attending to such water station could also attend to the bridge. I wish all the stations designated with a star, and the pre-pay stations also called flag stations, where a train is flagged. During some of the summer months we may have no passengers to get on or off at probably one-third of the places mentioned, and every time a train stops and starts it entails an expense to the Company. Of course some of the places will build up and offer sufficient business to warrant making regular stops, but until that is determined we should keep our hands free to stop or not as business offered might warrant.

"You may arrange for constructing, by means of piles and bridging, the shore of the draw-bridge at Loggy Bayou, as long as the cost of same seems to be very much less than iron and concrete pillars would cost which probably can be put in later on to better advantage. In fact I figure to get down there in the future, and have figures for accomplishing certain results at a considerable less figure than appears from your estimates. Six dollars per yard for concrete is in my opinion excessive. I wish I could get this wire business so shaped as to allow me to get off from here, as it appears that I might do considerable good if I could be south.

"I notice that you believe that the steel span bridge across Coushatta Bayou could be left out of Mr. Otis' estimate. While I am not sure as to the necessity of having that 120 foot span, nor that the construction of the spans themselves is correct, that it should be a steel bridge and should rest on iron and concrete, of that I am convinced. When a drawing is once made it is easy to make an extra blue print. I take it that Mr. Otis has made drawings of this steel bridge, such drawings to show the girder work, and also showing them in position with the necessary supports.

I cannot today go into the details of your estimates, but will take up the item of six freight and passenger depots, 20′ by 40′ each. I believe that a station about 15′ x 40′ would be necessary at Knox Point and probably East Point; at Nagenta, Buck Hall and Crichton simply a platform on which the planters could place cotton in the cotton season, and would also serve as a flag station for passengers; Des Arc, Elm Grove and Moon

Lake as well as Curtis and Bossier might probably do without stations, simply a platform on one end of which might be a frame building, as is used by the Illinois Central Railway Company, being 8 feet wide by probably 20 feet long, divided centrally with a plain seat inside. This will simply serve as a shelter during a rainy spell, and will be quite sufficient in places where we cannot afford to keep an agent. Any of those places that prove in future to offer sufficient business, and thereby earn better accommodations, they can have them later. I was of the opinion that the railroad track scale had been ordered by Mr. Rose. Same should not cost $1100.00.

"An engine round house if made of wood should not cost $1500. Though I figure it would be best to make it out of brick. In estimate and description of work, I notice Wimbish grading at 12½ c/yd., I was of the opinion that we had a contract with that gentleman of 11c/yd. I notice that in section 30 to Coushatta you are to pay 12 c for grading, the amount is given as amounting to $24,168.00; as that part of the grading is in the hills principally it does seem too remarkably high. I do not know what you have in the contract as to the 500 ft. free haul, by which is meant, if there is cut out of the hill a certain number of cubic yards and dropped into the adjoining hollow, then the measuring of the hill alone is paid for and not for the dump, though it may serve our purpose, whenever it is contracted for that all measurements count cutting as well as filling, then the contractors make lower prices for the work. Clearing at $22.50 per acre is more than it should have been. Trestling and piling I do not know sufficient about to judge of it. Track-laying and surfacing should not cost $500.00 per mile. Nor should we do any grading, at least not to amount to anything, for side tracks. I notice you figure cattle guards to cost $7.00 apiece; that is undoubtedly too much. We should not pay for switch ties by board measurement, $12.50 per hour should engineering cost us $10,000.00."

Bill believed that one of his agents was too expensive and that the agent should be replaced by a salaried inspector:

"The Rookery
3/22/98

Dear Clarence:

Now to learn that Mr. Bell resided 49 days at the Southern Hotel, St. Louis at our expense is a statement quite refreshing, and to see the footnote of Mr. Rose at the bottom, recommending that in addition to $226.75 our Company place an additional sum, touches almost upon the ridiculous. The expense bill alone amounts to nearly 40 dollars per car, while I am convinced that I could have a mechanic with our specifications in hand, examine each car as it left the shop at a cost not to exceed $2.00 each. It has all the appearance of having been set as a trap, and I would like to know how matters are progressing, and what you consider the value of Mr. Rose at this time."

Running short of money, Bill wrote:

"The Waldorf
Fifth Avenue and 33rd Street
New York
February 27, 1898

My dear Ellerbe:

"Arrived here this morning on telegraph call from Mess. Gates and Gary, later our Company Chief counsel; and it appears as if the Wire Scheme would not go through! It is not yet final, but it looks as if the financial underwriters did not consider this a favorable time, as all Stocks are declining owing to war talks with Spain. Tomorrow (the) 28th, most of the options expire, and if not accepted, I doubt if they could be renewed.

"So it seems as if I could not proceed with additional undertakings till I can see a way to obtain more cash. It seems as if present enterprises would over run estimates; how much I would like to learn from you. Let us be cautious so as not to spend moneys if we can not see either prompt returns or absolute necessity for it.

More soon."

Bill's shrewdness about business and his frugality are revealed in the next letter:

"The Rookery
3/29/98

Dear Clarence:

The contract that came from the Missouri Car & Foundry Company was $545. for box cars without brakes, or $580.00 with brakes; and $425.00 for flat cars without, or $460.00 with the Westinghouse brakes. The contract of the Westinghouse people with us or rather our agreement to use no other brake but the Westinghouse brake, entitles us to a $32.00 price for each brake for the Westinghouse works, add to which $1 for freight is $33.00 delivered to St. Louis. There is then $2 left between what it would cost us in paying Westinghouse, or what it would cost to pay in full to the Missouri Car & Foundry Company."

Bill was concerned about the effects of the pending war with Spain. He wrote this letter cautioning against setting shipping rates too low:

"The Rookery
4/8/98

Dear Clarence:

Regarding yours of the 6th you give a pretty full explanation of your view regarding the agreement with Stone Lumber Company. Following the line on the map of the State of Louisiana it does look as if the shipment of lumber from Shreveport to New Orleans via Vicksburg is a very round-about way of shipping. Your opening statement on this subject, however, is very broad, and I hope will not cause us any trouble in the future. You say: "I will guarantee that the rate from Shreveport to New Orleans is 15 cents." Now if by a combination of the various railroads rates should be advanced, as for instance, should the trouble with Spain lead finally to an actual conflict, the roads might be so deluged hauling stuff to the gulf that they might advance rates on undesirable freight which pays a low rate. Then if we had guaranteed 15 cents, and the rate were advanced to 18 cents or 20 cents, we might be in an uncomfortable position. Understand, however, we are having directors' meeting and executive meetings daily. I cannot go into this traffic question as thoroughly as I ought to."

An agent of Bill's had bought faulty cars from the Missouri Car & Foundry. The Mr. Bixby mentioned in the letter later had a building at Washington University named after him.

"Regarding the cars, you have probably misunderstood regarding my position. While it is true I would like to see the cars received by you and in good condition before I pay for them still they are due and payable when they have been turned over to the railroad people and especially when accepted by our inspector. This matter of inspecting the cars has been very unsatisfactory, and I have talked very plainly with Mr. Rose regarding the subject. Afterwards he stated that he would go down to St. Louis at his own expense and would interview Mr. Bixby of the car company regarding the matter. I do not know that it will do us much good, however, if the cars are otherwise in thoroughly firstclass shape and the only difficulty that the dried wood in the box itself swelled up when it came into a rain; this may not be so very serious."

Bill had decided to build his line to Natchez. Later, after this letter, he changed his mind.

"Consolidated Steel & Wire Co.
May 7, 1898

Mess. McIlvried and Ellerbe

Edenborn at Black Lake Louisiana

Gents:

And I think you should build from Coushatta or Campti toward Natchez. If you go to Alexandria it leaves for the Junction City or some other the immense lumber traffic between Vicksburg, Louisiana & Pacific Rd and Natchez. If you get there first, it is not likely that anyone will build from Campti to Alexandria and no one will likely come into that territory if you are once there. Get an early connection with the Iron Mountain. You will get the best rates from them. They charge 15¢ from Alexandria to St. Louis, 22¢ to Kansas City and Chicago, 34¢ to Denver."

Yours in haste,

William Edenborn"

In the following letter Bill did a cost breakdown and discussed how he intended to raise additional funds:

"The Rookery
May 10, 1898

Mess. McIlvried and Ellerbe

Gents:

I shall hereafter address Shreveport & Red River Valley Railway Company when I desire to give information to you both. I have mailed you today check for $10,000 which brings the amount advanced to

Ocean going Steam Ship

Louisiana Central Construction Company over $325,000.00 I have probably enough more to complete the Railroad to Coushatta. From Coushatta to Campti it will cost about Grading Ties and right-of-way $63,000; Rails $1000; machines $40,000; Fastenings $4,000 which equals $108,000. I can probably get a loan from A. Clifford of $50,000 and can perhaps dispose of some bonds and stocks and collect some notes to cover the balance. The difficulty with my case is that I ought to purchase large bodies of Timberland so as to get benefit of advance which the road will be sure to bring."

When he found evidence of surface natural gas he wrote:

"From the blueprint it appears that you have option on land of Prudhomme adjoining Fleming Lake. I think that place has on it that cold Water Spring near Alpha, said to emit also Natural Gas; That I should like Peter to examine, or both of you, and if you advise I will take it."

Bill wanted to locate a lumber mill on one of his railroad lines. On May 11, 1898 he wrote:

"H.A. Long, Esquire
Lumber-Kansas City, Missouri

Dear Sir:

Your Mr. Mansfield of St. Louis requests me to see or address you regarding Pine Lands.

I have just put in operation the Shreveport and Red River Valley Railway-Shreveport to Coushatta. I have options on about 12,000 acres of Pine Lands and am looking for some strong concern to locate Saw Mill on the Railroad. I learn that yourself and associates hold options closest mine and also that you intend to locate a Mill on the Shreveport & Red River Valley Railway. It occurred to me that it might be well to come to some mutual understanding in the matter, and avoid misunderstanding and friction. If you intend to operate mills on my Railroad from which to send to Market the product of the Pine Lands from north of Lake Bistineau to or below Coushatta, then I would have no object in getting other Saw Mills for that Territory and might arrange to turn over any options to you. I have excellent Mill sites with Lakes for holding logs if desired, good quality of water for Steam, and would like to have you call at the Railroad office in Shreveport to take you down the Line or write to me fully and promptly care of American Steel and Wire Company, St. Louis, Missouri.

Respectfully yours,

William Edenborn"

When the Loggy Bayou Bridge had been completed, Bill and Sarah took a speed run on the train for about 40 miles, between Coushatta and Shreveport. The train seemed to have the speed of a projectile as it barrelled through the dense virgin pine and the dark shadowed cypress giants. To Sarah and Bill this little railroad was like a new home and they planned to have their own railroad car. They planted brown-eyed Susans along the right-of-way, so much like home did the new railroad seem to be as a Degas painting on the wall.

But it wasn't all pleasure, for Bill still found that his competitors schemed against him. Edson, with whom Bill had tried to be friends, had gone so far as to build a Spite House in Shreveport, and there had managed to examine records of Bill's purchases, land activities and even his hiring procedures.

Bill continued to need cheap lumber. On May 11, 1898 he wrote:

"Gentlemen:

Your map of Louisiana will show, and you are probably acquainted with, the location of Lake Bodean and Lake Bistineau on the northern borders of which is found that fine grained, nearly all heart pine which has become known so favorably as "Lake Bodean Pine". As no railroad was ever built along Lake Bistineau a virgin forest of pine is found bordering its shores which extends fully 20 miles northward from the lake. The outlet of Lake Bistineau is called Foggy Bayou which empties into Red River about 30 miles below Shreveport.

"The Shreveport and Red River Cally Railway Company is now ready to take care of traffic between the points named. A saw mill on its line at a lower end of the Lake Bistineau is considered especially fine location. Many a settler owns 80 or 160 acres on the lake with very little of it cleared, and during most of the year saw logs could be secured at the mill at minimum cost. Close to the Lake the country is undulating, the drainage toward the Lake; while some distance northward the topography is favorable for inexpensive tram for tracks. We have options on over 4000 acres (timber) at from $1.50 to $2.50 per acre; and plenty more can be obtained. We will probably be able to offer Shreveport tariff rates, and if you contemplate an extension of your lumber business, we shall be pleased to correspond or meet you at Shreveport. A mill of 30 to 50000 daily capacity at least is desired. A prompt response would oblige. "

Yours truly,

William Edenborn
#925 The Rookery

Then Winn Parish demanded a rail line from Winnfield through Black Lake in exchange for tax subsidies. Bill wrote:

"Gentlemen:

I therefore take the sketch sent by Mr. Ellerbe, locate on it as closely as I can Flemming Lake, draw a dotted line from it through a narrow place in Black Lake and thence through the same narrow place in Saline Lake, through which Mr. Ellerbe showed what he considered a feasible route. If Mr. Ellerbe is correct that it is feasible to get with a tram road through Saline Lake, then I consider it also feasible to run with the tram road through Black Lake, because Mr. Whitaker states that in several places Black Lake was only about 300 feet wide, which would require a trestle, not necessarily very expensive. I take it that this lake is not very serious to get across, because in the map Mr. Ellerbe mailed to me a few days ago indicating in red pencil marks timber options take, several forty acre pieces are marked right inside of Black Lake. I believe that all these lakes have simply now a stream running through the center of same. Of course I figure that from Lake Flemming to Winnfield an inexpensive road of probably forty miles, section rails could be built for close to $100,000.00. Of course it would depend upon a close examination by Mr. Otis as to the feasibility of getting through the lakes and the hills. I have simply taken it as probably to be accomplished noticing that Mr. Ellerbe outlines part of this road. In fact the larger part of it through the same territory. I would not object to this road having one per cent grade. I would not necessarily agree to come with this road into Winnfield itself, but probably a point midway between Winnfield and Carthage. Probably Winn Parish would vote us a subsidy. We need not tell them exactly the route we intend to take and probably the previously voted subsidy could readily be set aside. With a subsidy voted and ten years taxes free, this projected road would be one of the best feeders our main line could get. FromCampti to Alexandria would probably remain our territory for some years to come and we might extend to Alexandria later."

While he planned the rail line to Winnfield, he decided not to enter the city proper because of land costs. He wrote:

"To the Winnfield people we might state that we would only bind ourselves to come within three miles of Winnfield; that we might come right into Winnfield, but would not promise to do so, because we would study up the engineering question and might find it unsuitable to do better than come within two or three miles of their town. Of course if necessary, two or three of the influential people could be promised a portion of the town lots should we find it expedient to lay out a town in the vicinity centrally between Winnfield and Carthage. This I believe will give you at this time all I have to say on the question under consideration, and as further information comes to me, will keep you posted."

Trying to reduce a tax assessment, Bill wrote:

"Regarding the taxing of our road at $4,000.00 per mile I believe that we should make an effort to have this reduced. I do not understand how we should get on the wrong track repeatedly by statements volunteered by Mr. Dortch. I am convinced that Kansas City will not pay taxes at a rate of $8500.00 per mile."

Then there were problems with getting the right direction for his rail lines. Bill wrote:

> 'If we can't get a feasible road from Grappes Bluff to or near Winnfield with future extension to Trinity, then we must try to get down to St. Maurice, get hold of some good tracts of long leaf Pine and get someone to tram from there or from Montgomery directly northward toward Bienville. If I hear further from my letters to Yellow Pine lumber men, I will let you know. I'm sorry to learn of Bridge 22 burning. I expect to come down soon with Mrs. Edenborn, meanwhile remain.
>
> Yours truly,
>
> William Edenborn"

Then his competitors, The Cotton Belt, began foul play. On June 14, 1898, Bill wrote from the Rookery:

> "Dear Clarence:
>
> I notice that the Cotton Belt pursued their previous tactics of ditching a car when they feared we would proceed to cross their tracks. I would advise you to engage a detective at a stipulated sum to get information as to who gave the order to the switchman or brakeman to ditch the car. It should not be difficult to get information regarding the matter, and if obtained I would instruct you how to use that information."

Bill wanted all reports on his latest endeavor sent to him at the Rookery. He wrote:

> "Regarding the operation of the train, I should get a report of the earnings twice a week, once for the first 3 days and once for the last three days. I think the earnings should be divided one-half to the Construction Company for operating the trains and one-half as a fund to meet the interest. Probably you may consider this division a trifle different though I believe it is not far wrong."

The Cotton Belt struck again. On June 16, 1898 Bill wrote:

> "My dear Ellerbe:
>
> I think you should not hesitate to have known to businessmen the threats made by West; and when the Cotton Belt ditched a car to prevent us from placing our crossing, a little item in the paper should inform the Public that: 'The Valley Road tried to finish its Track but found that the Cotton Belt had a car off the track just where the tracks would cross—it was repetition of what happened last fall at end of Commerce near Cross Bayou. The officers of the Valley Road say these things of course are costly and annoying but will not deter the Valley Road from carrying out its contract with the City - while the Citizens realize the spitework so plainly displayed and will keep it in mind.' The Courts will probably sustain us, if not then the Citizens must speak out in unmistakable terms and put its ultimatum to the Cotton Belt either to grant us that strip or move their West track three feet more west.
>
> "It's got late - more probably tomorrow."

When he was asked to locate a station along a curve which he considered dangerous, Bill wrote:

> "Dear Clarence:
>
> I note your remarks regarding East Point town site. Mr. West and the citizens of East Point cannot in my opinion compel us to locate a station on a curve, and in order to avoid placing the station on a curve, we should go either North or South from that curve, and considering the conditions of the locality, height above low water, and engineer's recommendation, together with the accessibility of the site selected I doubt that Mr. West would succeed in compelling us to change locations. However, as the other party saw fit to recede from his original proposition, we may leave the quarrel to the parties down there to decide. Meanwhile, we should incur no further expenses, for erection of the depot & other improvements."

Bill attempted to use the Cotton Belt Railroad track in Shreveport. On June 21, 1898 he wrote:

"Dear Clarence:

Regarding the agreement you made with Edson, I will say that it was probably not wise for you to enter into the negotiations between Edson and the Committee and offering to waive all damages we might sustain, yet agreeing upon an enormously large figure to be paid to Edson for the right to run over a City street. The citizens of Shreveport must pay very little attention or you must have very little faith in their views of honor, if they do not carry out their contract with me to afford me facilities over street and batture to reach such places as the business of the road demands, without paying heavily for the privilege of using a City street to a competitive railroad; but while I am free in expressing my opinion on that subject, I realize at the same time that I gave you authority to make such an agreement with Mr. Edson as you considered best and having done so, I do not desire at this writing to deviate from the agreement you made. Meanwhile, I notice that Edson is not coming to Shreveport to settle the matter promptly with you, while you write that you expect the Supreme Court's decision at any minute. Probably Mr. Edson thinks if the Supreme Court decides in his favor he would say the Directors of his Company would not allow us the track on Commerce Street, but if such a decision is in our favor he would probably then take the amount of money you offered instead of the amount of the jury's reward."

Remember the faulty railroad cars that Bill complained about earlier? How he had to file a complaint about them. In this letter, he also revealed his sense of humor. He presented this claim to Bill Bixby who later became the first President of American Car and Foundry, Railroad Car Trust.

"William K. Bixby, Vice President and General Manager
Missouri Car and Foundry Co. - St. Louis

Dear Sir:

Yours of 21st received. I called at your office on that date; as you were absent I left with your clerk copy of the report regarding those cars. In addition to Inspector Captain Moncure's report, I have statements by Car inspectors and Car builders from various Southern Railroads notarially acknowledge which reiterate Capt. Moncure's findings. Your request to return the Cars after we have paid you over $28,000.00 is about as 'sappy' as most of the Lumber in those Cars. No, the Cars are needed, but they are *in evidence*; they tell the tale as to the quality of Lumber; the Specifications are also in evidence. We shall not pay the $1400.00 you claim because our Claim against you being now formulated will be much in excess of that figure."

When planning where to locate train stations Bill wrote on June 24, 1898:

"Dear Clarence:

Regarding the depot buildings, it seems to me that we should put up such buildings at Coushatta, another at Curtis, one probably at Knox Point, and I should recommend waiting furthur developments before deciding on station building at East Point. This matter is perhaps not so much for me to judge as for yourself. Probably a station at Coushatta should be first put up."

American Steel & Wire Co.
Havemeyer Building
New York
June 27, 1898

Mess. McIlvried and Ellerbe:

"After visiting the Cleveland Wire Mills, I arrived here this morning. Shall not likely get any letters from you for a day or two, but believe you have my views on the several matters pending. Re: Ballast: The Roadmaster of Cleveland & Alton Railroad told me that McAdam (broken Stone) was the best Ballast, next Furnace Slay, next burned Gumbo - of the right kind - next Gravel-Sand, only if nothing else available! He said whatever the Alton put down, McAdam or Slay, they put on top a dressing of Cynders; they would place cars for receiving Cynders at their round houses and at Factories where they could get them free. Now we can't get McAdam or Slay now nor Cynders very soon - So I think we should try Oil on the Sand.

"Train now in operation. Statement received show average daily earnings $50.00 practically. We have got lately all passenger traffic between Shreveport and Coushatta and while we may add some little to freight and perhaps to passengers - yet the earnings indicate what we may expect any time outside of the Cotton season. I have had estimates as to earnings of the Road Shreveport to Coushatta from several sources. I considered the estimates of Mr. Clarence Ellerbe as probably most reliable because made after carefully investigating the Subject. Our earnings fall short of expectations. That we may think to extend the Railroad to get into the Lumber regions and to get where considerable more Cotton can be controlled, is going into a different field; we should see now if that road Shreveport to Coushatta will pay on the Bonds and a something on the Stock as it was originally figured, - not forgetting to allow a due portion for renewals of ties, Bridges, etc. Dr. Houck of Cape Girardeau road told me, it did not pay to haul a train that did not bring $100.00/day. It is true his road is twice as long as ours, but then we haul a train each way and get only $50.00! Passes: In comparing what I see traveling on other Roads with our road: I find that our Conductor does not even require a passenger to *show a pass*! It somehow seems a foregone conclusion that the principal people along the route have a pass! I am of the opinion that we have nearly doubled the value of the Land. I believe we save the Cotton grower an average of a dollar a bale on Shrinkage in weight deterioration, interest, and cartage; we save the planter on his Cotton seed, etc. On other freight, we save the man from Coushatta about 75% in time and in fare to Shreveport and back - and that Planter will make with reasonable management 3 to 4 times as much on his investment than we do - under such conditions, it behooves us to be *just* before we are *generous*! If you have to issue a pass do so for a round trip, so it ends there - do not issue annuals to *anyone* without my consent. If one Planter rides free and another has to pay, the one that has to pay will ship his Cotton by Boat in preference. Trip passes issued should bear a number, date and name of passenger and the Conductor must report such name or number or both! If annuals are out - don't hesitate to tell that train earnings don't warrant passes.

"Meanwhile, I am convinced that I must build to Campti and perhaps to a place on the Saline 2 miles above St. Maurice, or else Fordyce (promoter of the Cotton Belt) and Case may get ahead of me."

Then low-priced American Steel & Wire Stocks, due to the Cuban War caused a shortage of funds. Bill wrote:

"Dear William Whitaker:

Re: Timberlands. I don't think I shall be able to buy them - I am trying to raise the Wind to build to Campti - hope to succeed. Naturally, in these times it is hard to say what turn may set in in the Stockmarket and as you say, the matter can be talked over when I get down there - say July 8th to 18th - can't tell exactly."

Bill and Sarah had sailed on the ocean going steamship, The New Orleans which plied waters between New Orleans and New York City. They had considered bidding on her but she was too large for their requirements.

Bill decided to buy the *Grapevine*, a famous steamboat that would have had the capability of transporting cotton between New Orleans and St. Louis. On July 5, 1898 he wrote:

"Re: Steamboat, I do not learn of many suitable Boats being for sale; probably the two boats that were used on the Monongahela which Captain Moran was to get price on (about $13,000 to 14,000 each) might be suitable; one of them is about three years old. The Steamer Grapevine lying at Carondelet (a suburb), I looked at yesterday; she is an old Boat but received a New Hull last winter, and seems in good condition. She was formerly called Woodford or Woodson, her last service on the White River and some on Arkansas River in the interest of Iron Mountain Railroad. When she required a new Hull and general overhauling last fall, Captain W. H. Grapevine, who is in charge of Car transfer, ferry for

Railroad Transfer Boat

passengers and teams, was requested by his men to give the Boat his name which was done. The Iron Mountain Railroad had another Boat: *Steamer Missouri* which has just received new Boilers - and as the Boats on Arkansas and White River have quit cutting rates, learning that the Railroad could with one or two free lance boats make such rates as to wipe out all profits - the Iron Mountain Railroad desires to sell the Grapevine. They ask $9,000, don't know if that can be shaded. I think she might suit our purposes; I take it that we must have a Boat to comply with our contract, and probably as an argument to get our share of out bound Cotton. As soon as we have bought a Steamer, we can then inform Commissioner of Shreveport Board of Trade that we are prepared to ship to say Campti by rail and then by Steamer. The Grapevine has good passenger Cabin accommodations - rather more I think than necessary. If there was more space for freight, I should like it better, but I suppose the space is as much as boats of say 250 ton burden usually have. I was favorably impressed with the Grapevine."

Bill tried to buy land for his railroad and to avoid a confrontation with another competitor, the Buchanan Railroad. He wrote:

"I note that you expect price on Texas and Pacific Lands between Black and Saline Lakes soon. I note also that probably the Buchanan Sibley combination intend to build into that territory - if so we had better secure the Timberlands north of Lake Iatt, said to be fine; if Buchanan et. al. build in between the Lakes they may find that by that time we are past St. Maurice and they then would probably connect with our road and be a feeder to us while, if we secure the Timber between The Lakes, then they will stay on Winn Parish territory and have an open chance to extend later to any territory left open by us. I believe I can discern the one that is hatching these schemes and puts them before our opponents; still building Railroad costs lots of money!

"I hope you will get your Rubieaux deal through. Re: Trestle to Press along cross Bayou, it's well to get figures and will decide when I come. Has not Hepburn finished laying Crossings to McKellar property and mapped out terminals there *and left*? Are you not yet running into Coushatta?"

When he anticipated trouble with promoters, Bill wrote on July 6, 1898:

Dear Clarence:

"I note figures you name on Cotton. I think we should have no difficulty to get our percent in the Pool. The Railroad Agents may leave us out, but the Board of Trade of Shreveport is also party to the agreement; and if we inform them that we are ready to take freight to New Orleans via rail and Steamer, they can ask the Railroad Agents to add our name to the agreement; and the simple demand that that be done will most likely settle the matter. Would it be wise to ask the Valley Steamer Line to name us prices on Cotton from Campti to New Orleans?"

Bill was tempted to build to Natchez.

July 11, 1898
Natchez Savings Bank

Dear Sir:

"Natchez is booming, and in building and business, going up now: New Bank, Two new Cotton Warehouses, One new extra large Merchandise Warehouse, Several stores and residences, our new electric plant, etc., etc. Hope soon to hear from you."

Yours Respectfully,

Abe Moses

Bill offered $7,500 for Grapevine, Eads Iron Works wanted $9,000.

"McIlvried that offer we made yesterday of $2,000 for Mogul Engine is accepted. We also offered $7,500 for Steamer Grapevine but seemingly not accepted - they ask $9,000. McIlvried wires our offer for Grapevine referred to New York officials (Iron Mountain) will take a week. Note letter from Abe Moses - have not had time to answer. Will be in St. Louis in the morning 14th. McIlvried will probably leave for Shreveport a day before me. Never heard from that London party about his Long Leaf Pine lands between Black and Saline."

Yours,

Wm. Edenborn

The London party was a Mr. Easton who owned critical lands between Lakes.

Bill's land agent, Poindexter Dunn, offered advice on strategy with competing promoter railroads.

August 15, 1898
Mr. William Edenborn

Dear Sir:

"Since our ride together from Coushatta to Campti a week ago, I have thought much over the matters that you spoke of in connection with the affair with Edson and the Cotton Belt, and its possible influence upon the interests of your road in the future. The total defeat Edson has sustained will rankle in the mind of a man of his disposition; and I am of the opinion that it may be important for you to consider of and take such measures as may seem best to counteract unfriendly movements from that quarter, and to fortify and reinforce yourself as much as possible along the lines of your road as built and projected. I therefore venture to suggest some interferences that may be anticipated and some movements that may strengthen your position.

"Beardsly is getting money and is pushing his road southward towards Alexandria through Winn, and his connection to the northward with the Cotton Belt about completed.

"It is evident he has, at least, the moral backing of the Cotton Belt. The Buchanan and Long Bell people have about $400,000 to $500,000 invested in railroads and sawmills, and are now investing $100,000 more in Yellow Pine plants south of Sibly; and, as I am informed by their timber gent, they are going to occupy the territory between Black Lake and Saline and thence into Winn Parish. Their roads are connected with the Cotton Belt, but I do not know if there is as yet any sort of alliance. The Henderson Brown-Junction City road, it is now said will at once commence building south through the Long Leaf Pine toward Red River. This also has a northern connection with the Missouri Pacific System. All these projects are moving in converging lines toward Red River through the great Long Leaf Pine Lumber region east of the river; and if they all secure the support of Gould System to an extent sufficient to bring them under that control, it is manifest that they can make a drain from territory that should be tributary to your road.

"I will here remark that in all that I ever wrote and said in the beginning and later progress of the work of promoting this enterprise, I considered and treated the lumber traffic to be drawn from the great Long Leaf Pine forests of Winn, Grant, Catahoula, and other Parishes as the very backbone of the enterprise. I advised the locating of the road in Red River Valley from Shreveport to St. Maurice, (the edge of the Long Leaf Pine) so as to secure subsidies and a cotton traffic along that part of the line where no lumber traffic of moment could be expected. The movement of all these newly projected roads into that great undeveloped lumber region shows how valuable and important both Railroad and Lumberman consider it.

"Some sort of scheme should, in my humble opinion, be devised and promptly and strongly pushed to secure a large share of the stumpage of that great forest of the Long Leaf Pine. If it is not practicable for you to build into it and secure it, then seek an alliance with some great lumber and timber sydicate who will secure, or have, the timber, and who can and will advance the moving to carry your road on from Campti to that great lumber region.

"These things secured there can be no question as to the success of your road. If the Gould People intend to persist in their hostility, as I here anticipate, by the means above suggested, your only counter movement is by a speedy occupancy and development of as much as possible of that lumber region.

"I would in this connection sound the Buchanan, Long-Bell Yellow Pine people for an alliance and if possible detach them from the Gould Alliance. I have no doubt that a combination can be formed with some great lumber syndicate free from Gould influence.

"If you lop off the lumber end of the project and determine to have mere valley railroad, then it seems to me you had better consider the feasibility of expanding to the eager overtures of the City and Parish of Natchitoches, cross the river below Grand Echore, go down through the old Cane River crossing via Natchitoches and secure, if possible, all of Welch's franchises and subsidies; get as much more as possible on the West side of the river and go on to the Mississippi at Baton Rouge or New Orleans. Of the two schemes I prefer the former. One or the other must be done to achieve success. To halt in the middle of this valley will be fatal, I fear."

Faithfully yours,

Poindexter Dunn

August 24, 1898
Shreveport and Red River Valley Railway Company

"I trust when this reaches you, you may have word from the New Orleans parties regarding the timber and lime stone quarries (Coochie-Brake). I think if I have secured that tract and build to St. Maurice, it will take the wind out of the sails of several promoters. I hope you will leave no stone unturned to push the construction of the Road to Campti, because if we want to go beyond Campti, we must do it during the Winter before cotton is planted, owing to the open fields through which the road is to run."

Yours truly,

Wm. Edenborn

August 27, 1898

P. McIlvried

"Re: Station houses, you will please keep them at some distance from inflammable buildings such as Gins and Seed houses, but not insure them. I will take the insurance risk, by having a fund created based on what premiums for full insurance would be and crediting this fund with it. -Though I don't want a very elaborate system of accounting for it started.

"I intend to leave for Cleveland tonight and may go to New York and other places. Hope to learn of your getting off a little and come east."

Yours,

Wm. Edenborn

A La. lawyer suggests land subsidy for Bill.

August 29, 1898

Mr. Abe Moses

"Absence prevented sooner reply to last favor. I enclose you a statement of 51,867.00 acres of Land in this (Catahoula) Parish owned by Tensas Basin Levee Board on probable route of Railroad from Campti to Jonesville. These Lands are mostly pine lands off of which none has ever been cut. They have no fixed price, varying from 60¢ to $2.00 per acre. Sale of same will close after September 27th."

Yours respectfully,

J.B. Boatner

Bill answered Abe Moses' overture to build to Natchez instead of Alexandria.

September 5, 1898

Abe Moses

"Yours of 1st to hand also enclosures of map from Mr. Boatner. I accept your kind offer to procure that Map - which, if I understand correctly, is to show exact location of Lands - *all Lands* - of the Levee Board, and from which they offer 1000 acres at 8¢ for each mile of railroad built within their District. By the way the State of Arkansas lately offered 2000 acres free per mile. Now I was of the opinion that a Levee Board had only Swamp on Lands subject to overflow to offer and I notice these Lands named by Boatner are in *Undulating Pine Hills*! Principally for this reason I ask for the Map: because such Hills may offer Engineering difficulties, grading necessitating another location to reach a given point."

Yours truly,

Wm. Edenborn

Gents:

"Note Abe Moses' letter and enclosures! I had Murray make a sketch on back of this to show location of the Lands. They are owned by Tensas Levee Trustees and can be had by intending Railroad for 8¢ per acre. If their real value is $1.08, then the 51,867 acres would be equal to a subsidy of that many dollars and probably Catahoula Parish would also vote a tax subsidy. I made a Sale of Wire Stock and sold also my White Cliffs interest thereby increasing available cash by $70,000. This for your guidance. *Confidential*!"

Edenborn

Bill tried to buy enough land to keep the railroad busy and the cars loaded.

August 30, 1898

Gentlemen:

"I have again read report on Timberlands and will say that after the examination by C. Ellerbe and further examinations by Mr. Love or others, you viz: P. McIlvried and C. Ellerbe, can best determine by actual figures how much Lumber, a purchase of say 20,000 to 30,000 acres, would give to our railroad per annum through a series of years - 10 or 12. You are to agree on a plan of procedure and make your joint recommendations. To give you my views as to financial possibilities will say that I consider *Construction* from Shreveport to Coushatta as closed (virtually); cost of Construction to Campti say $110,000; or say Coushatta to St. Maurice $180,000 - we may do it for $150,000 - or $3,000 for rails, fastenings, and ties may do it! If so, I would have say $150,000 to $175,000 available, for extension to Winnfield and Timberlands; while I strictly enjoin you from allowing this information to become known to others, I give it to enable you to form your recommedations. I further intend to dispose of some small securities and some property in St. Louis which may increase the above figure $50,000 but don't want to rely on that too much."

Yours truly,

Wm. Edenborn

Bill tried to beat his competitors to Alexandria.

August 30, 1898

Gents:

"Considering the Timberlands, it strikes me that we have fought shy of *Grant* Parish because Goulds own much of the Timberlands. Now if we could secure Lands say between Saline Lake and Iatt Lake and Creek -and

Winnfield north and Nantaches Creek south, it would keep those railroads north of us from going to Alexandria, and if we built our railroad to Winnfield, they might keep north of us or go east of us or connect with us; by this leaving that strip between the Lakes (Black Lake and Saline Lake) if anyone came down through Ward 2, they would connect with us as a feeder. Getting sufficient to control the territory described would put us in a commanding position, I think. What say you?"

Yours,

Edenborn

Bill raised more construction money.

September 2, 1898

"I had rather a tough day of it, but am pleased to report that I have received Certified Check for my White Cliffs interest. I had to make a concession to the Kelleys, but I have $50,000 back, which is very good considering the bungling work of the Lawyer - Judge Gary's partner."

Yours,

Wm. Edenborn

September 5, 1898

"Arrived here this morning. Tried to come to an agreement with Illinois Southern Company to roll all or at least one day's product of rails and all angle bars without success. Finally appealed to Mr. Gates and told him that I would inform him in confidence that Carnegie Steel Company had lowered their price and agreed prompt shipment. He said he knew they could not - though they like other manufacturers might promise."

"Re: Cotton Belt. When I was down at Shreveport last, I told Mr. McIlvried that I saw a gang of say 5 men with Engineering tools go into their Spite House! Well we are in it to stay, and I propose to fight it out. What say you?"

Yours,

Wm. Edenborn

The St. Louis Mill had been closed.

September 9, 1898

"I see from yours of 7th that it turned very cool down at Shreveport. The heat of course is practically over. Trust this will find you quite well, being more acclimated. Our Directors voted to start St. Louis Mill and I will have to be there a while. But I will get to Shreveport as soon as possible, and Mrs. Edenborn insists that she will go with me."

Yours truly,

Wm. Edenborn

A new mail contract was proposed.

September 13, 1898

"Re: Mail. Will retain the P. M. letter etc. as I might get to see Mr. McIlvried. Meanwhile I think we might accept of the offer of $1,500 per annum up to June 30, 1900, and then have weighings to determine what we are entitled to. Also we will be at a different terminal in 1900."

Yours,

Wm. Edenborn

Bill wrote this letter regarding diverting the road to Natchez.

September 16, 1898

Hon. J.B. Boatner
Attorney at Law
Harrisonburg, La.

Dear Sir:

"Returning from New York I find among accumulated mail, yours of 6th inst. Thanks for your kindness and willingness to work for the development of your Section of Louisiana. However, I am not ready by any means to undertake the construction of a railroad toward Natchez. Mr. Abe Moses of that City, apparently an indefatigable worker, is endeavoring to have the Shreveport and Red River Valley Railway extend from Campti to Natchez. I am operating from Coushatta to Shreveport, 45 miles, am constructing from Coushatta to Campti say 18 miles, and if funds hold out may extend to St. Maurice, but beyond that??? I hope times may change, but as yet I have not been able to induce any Capitalist to join me. I shall address you again later, meanwhile remain,"

Yours truly

Wm. Edenborn

Bill was offered stock in the Shreveport Bank.

September 16, 1898

C. Ellerbe

Dear Sir:

Answering your various letters: the Shreveport Bank Stock deserves most serious consideration. Probably 52% to 55% of the Stock would give us considerable standing and pull. What I don't know, is the reason those 'Grocers' who tried once to get me interested in another Bank and found me, not averse, but unable to shoulder the whole enterprise, did not say a word when the Levy stock was for sale; took what they wanted and let balance go to 2 or 3 outsiders. My own information given me in Shreveport was that Captain Youree, while President of the Bank, had but little Stock in it! Now should it be that someone thought the Bank had more profit in it, found out different, or probably found its *assets* or *loans* too *hazardous* and wants to get from under? You say you know it paid 8% last year, and you were told it earned 22%. Now surely those that bought the Levy stock know what the Bank paid for a series of years back. I feel inclined to purchase the Stock; however, I know so little about banking that I would have to trust very much to Mr. Hamilton, probably might have the privilege of naming Cashier or Bookkeeper. I can not get down there till toward end of next week at best. Meanwhile, you may get further information, and if favorable, I may take the Stock. Keep me posted c/o St. Louis office of Wire Company. Did not get letter in time to consult Mr. McIlvried. Returned here this a.m. Leave for St. Louis tonight.

"Re: Mail Contract: McIlvried says you intended writing to Mail Supervisor, Dallas to the effect that we would accept at rate of $1,500 per annum up to June 30th. 2,000 and thereafter we get whatever weighing would show we were entitled to. I agree to that. Will make attempt to have rail routed as requested. In fact Freight man of Illinois Southern Company says he will ship as requested if it costs no more."

Yours,

Wm. Edenborn

Bill planned a sawmill.

September 21, 1898

Gents:

"Had W.L. Whitaker yesterday. He says he will make a proposition - the last - for the good of the Shreveport and Red River Valley Railway. His plan is to obtain for us a Saw Mill of 75,000 feet per day located on our tracks at Bossier. Get a lot of Russell Log cars made at Detroit, capacity about 3,000 feet log measurement instead of 34 feet flat car with capacity of 5,000 feet to 6,000 feet. They cost say $175 each. Haul these cars say any place west of Campti at $2 or $2.25, and get outbound freight. This he claims will only be 1/5 of the logs but will pay us daily $100. The

Saw Mill would profit by selling slabs, etc. to Brick Yards and kindling, shavings, saw dust etc. in Shreveport. This plan he says will bring the scattered timber to a central point to our benefit. I told him to put in writing the plan and his recommendations. Am waiting."

Yours,

Wm. Edenborn

Bill tried to get a bridge subsidy for Grand Echore Bluffs.

September 28, 1898

Gents:

"A Subsidy of $100,000 payable during 20 years, is not so large an amount as it looks - yet $5,000 per annum is a 10% investment on $50,000, and I think whoever builds the Bridge takes risks that are hazardous and ought to pay 10% - I believe Natchitoches will vote us a 5 Mill. Subsidy and Ward 9 may vote us 5 or at least 2 1/2; and probably we could get that 4 M. track to Grand Echore free. Then we would have to run into Natchitoches. Of course if we extended to Alexandria, we might simply run Engine and one or two cars from main line to Natchitoches. Of course we would have to Contract 1st with responsible Bridge builder. The Bridge would not be of much value as a revenue producer for some time. Undoubtedly the Southern Railroad now at Greenville on the Mississippi River will want an outlet or connection with Texas, and a route from Greenville to Logansport, Louisiana via Monroe, and Grand Echore would lead them into the heart of Texas.

"The Illinois Central Railroad also wants Texas connections, and if we should build from Black River (Trinity) to St. Maurice, then a Bridge at Grand Echore would also prove a good thing. For the latter route viz. a Bridge at Colfax and through Ward 9 to Natchitoches or even at St. Maurice and through part of Ward 9 to Natchitoches would suit us as well."

Yours truly,

Wm. Edenborn

Bill tried to decide whether or not to cross the river at Grand Echore near Natchitoches.

November 17, 1898

Mr. Clarence Ellerbe:

"I had a conversation with Mr. Randolph yesterday. He told me that he hoped our road going to Alexandria would go down on the east side. I told him that we might do so if we found sufficient encouragement; that we however might contract with Natchitches for a bridge and go to Alexandria on the west side of Red River. He stated that he was convinced that Captain Jeffries, his father-in-law, would willingly donate five or seven hundred acres of land, and others on the east side of the river, in addition to voting tax, make donations of lands."

Bill tried to buy strategic lands that belonged to Mr. Easton of London. (same letter)

"Will attempt to see Mr. Huntington when in New York regarding Cheneyville connection.

"Will cross over to London and try to secure the Easton lands. Will then cross the channel into Germany where I expect to make a collection on patent sales heretofore effected, and return at earliest possible moment. Will write you from the east giving my European address and shall expect to hear often from you during the first two weeks after sailing from New York; beyond that time letters would probably not reach me any longer in Europe."

Yours truly,

Wm. Edenborn

October 17, 1898

Gentlemen:

"Re: Ties to be cut from Rubieaux Lake. I fear we may soon be in same position as Valley Southern & Pacific: unable to get ties! It may be difficult for us to get ties enough cut for our contemplated extension to, or nearly to, Clarence, La. I still believe we should reserve as many Cypress as possible for the alluvial lands and use the Oak for Hill lands. It appears that the principal difficulty now with Tie cutters is that the Oak Ties are so very much heavier than the Cypress. Now the Cape Girardeau (Dr. Houke's road) buy its Ties along the road at 15¢ for 6 x 8 x 8 on the right-of-way. I believe we now pay for 7 x 9 x 9 at 24¢ - now suppose we pay for 7 x 8 x 9 Oak say 20¢. I offer this only as a suggestion to be considered. I will say though: should we find our way clear to build to Colfax and thence through alluvial lands, all in cult. and no Cypress, we would be glad to still have that Rubieaux brake."

Yours,

Wm. Edenborn

Bill attempted to decide whether to build to Alexandria or east to Natchez. He tried to coerce subsidy aid from the Alexandria area.

October 30, 1898

Gents:

"Arrived here safely. Mr. Welch met me at Alexandria and rode as far as Bunkie. He asked what I wanted. I told him I was not quite prepared to answer; however, I thought I might decide shortly between Natchez and Alexandria if latter offered say:

For Railroad and Wagon Bridge: Rapides Parish Subsidy 2 ½ Mill
Wards, Alexandria (1,200 M) Pineville (300 M) & Ragonlia each 5 Mill
City of Alexandria & Pineville cash $30,000
City of Alexandria & Pineville each for terminals 30 acres
Guarantee right-of-way to Grant Parish Line
Colfax either $10,000 cash or $5,000 cash & 5 Mill subsidy
3 Wards in Grant each 5 Mill & right-of-way
Ward 5 in Winnfield 5 Mill right-of-way & terminals 400 acres

Welch will feel around and will probably know soon as to the chances to get the above. Probably we may use the Services of Welch and Sylveser if we want to get down there and what we want. I thought I would ask broadly, may have to modify some."

Yours,

Wm. Edenborn

Bill was distressed over high land prices.

November 18, 1898

Gents:

"Answered your telegram to buy at $1.65! Do not understand it. Mr. Ellerbe told me a week ago that those lands for which he offered 75¢ could be had as per Mr. Green's offer for $1.50. I requested Mr. Ellerbe to take them at that and he said he would do so promptly. I thought the matter closed, and therefore could not understand that the matter still open."

Lawyer Mr. Randolph was to obtain more right-of-way on the east side of the river. (same letter)

"Re: East side of Red River: If Mr. Randolph could pay a visit to his father-in-law by going to Coushatta and then by team down the East side and begin below St. Maurice to get Land subscriptions from planters, say 5% to 15% of their holdings to secure the railroad for their side and to avert the railroad crossing at Grand Echore going thence southward, that trip we could probably pay for as to success achieved; what do you think?"

Yours truly,

Wm. Edenborn

Bill was concerned about friction with the Cotton Belt (St. Louis & Southwestern).

November 19, 1898

Gentlemen::

"Have also seen C.D. Johnson, Manager, Frost Trigg Lumber Company, Equitable Building here (St. Louis) at N.W. corner of 6th & Locust! He says Mr. Frost has the money to put into a large Saw Mill plant and his proposition is about as follows: They to put a large Saw Mill consuming 12 to 15 cars at 5,000 feet log measure daily at 'Bossier'. They to take timberlands we own and options- (I claimed to hold extensive options - Easton, Violet, etc. - did we mention these?) - off our hands at what we paid and interest; and they to put our spurs from our railroad into the timber and bring their own Log cars to our railroad and pay us $5.00 per car for hauling to Bossier any distance up to 80 miles; and they to use their own Log cars of capacity same as used by Buchanan - he said 5,000 feet log measure but seemed disinclined that such be named as limit - that they would agree to put down a small Mill say at Campti or some suitable place where Long Leaf Pine is getatable to cut into beams, stringers, etc., but same to be hauled to Bossier same as log cars."

Yours truly,

Wm. Edenborn

Bill tried to get a subsidy for a major Red River Bridge from Pineville to Alexandria.

November 21, 1898

"Your second query: inasmuch as Alexandria and Pineville receive the biggest benefits of the road more than any other town in the Parish, it is no more than right that they should vote each a separate subsidy.

"Third: I believe we have put the cash subscription to be paid by Alexandria and Pineville low enough to command support. I would hold out very strongly for the cash, saying that practically everything is a promise to pay, but that we ought to have some cash, especially considering the fact that a good many tax aids have been voted for Railroad enterprises but none was ever paid. If absolutely necessary, have them reduce you to $10,000 instead of $15,000, but certainly we can use the argument that all they need do is to put with the Banks such values whereon the Bank will issue statement to us that they have received that for which they will pay us cash when first passenger train runs into Alexandria over our road. That gives them time in which really to make the payment.

"If they insist on the bridge being in the contract, you may agree that such bridge be for trains and wagons. A Toll Bridge in our favor with Shreveport rates is necessarily understood. Have Mr. Dunn canvass the situation thoroughly and if after all it is necessary to drop both cash and local subsidies in the two places named, you may agree to that after holding full council between yourselves.

"If I am still in New York, I would have you wire me first before you agree to drop as per your second and third queries. That several of the prominent planters from Grant Parish should favor a road down their side of the river goes without saying. To them such a road would be a Godsend, and I am convinced that the planters and land owners should make us land donations. I have written about that in a previous letter and trust Mr. Randolph will promptly act in the matter."

Yours truly

Wm. Edenborn

Bill called on Harahan. Harahan, Louisiana, near New Orleans, was named after him.

November 24, 1898

"While in Chicago, I called at V.P. Harahan's Illinois Central Railroad Office, but he was out of town. Talked some with his Chief Clerk: told him I had the Railroad at Campti from where I might continue to Alexandria and to Baton Rouge - or might leave Red River Valley near Campti and extend toward Natchez; that I believed the Illinois Central could bring much merchandise for Shreveport and Texas points, and if they desired or favored a connection with my Railroad and would promise freight to me that I would probably have much Lumber for them in return; that I was not ready to pledge myself positively, yet if I knew their wishes and intentions, I would duly consider them etc. He promised to lay the matter before Mr. Harahan with the request to write to me at Shreveport. So you will open letter and on my return from Europe may find copy at New York and then if necessary will call on him at Chicago."

November 27, 1898

"I got wet feet yesterday and today had a slight cold, and it being Sunday and rough out, have taken a rest and nursing. Wired you that I sail Tuesday morning a.m. on North Gen. Lloyd Steamer Friedrich der Grosse, have my plans laid for that Boat though have no tickets yet, and if anything needs attention will wait till Wednesday on the Teutonic. In the later case will wire you tomorrow. I believe that you have my views on how to proceed. First, that you vigorously, yet cautiously, secure all fairly good Timberlands ahead of probable extension of Buchanan or Beardsley because as we have not yet succeeded in getting even one Saw Mill located, we must ourselves secure the timber or our Railroad will suffer. You, of course, will endeavor to buy outright only the lowest priced lands and get options on others, unless you deem it necessary to close promptly. Try to get all possible information from Clerk or Deputy Clerk in Winnfield and Natchitoches on Deeds recorded or options recorded with values obtained on Timberlands by others."

Bill purchased interest in a Lumber Company later to be called "Urania". His partners were L.J. LaCroix and Henry Hardtner, the father of Southern Reforestration.

November 28, 1898

"I believe that in Mr. Raphiel you have a good man to secure timberlands from the natives, and I believe that with Mr. LaCroix you will have a reliable man for estimating stumpage.

"Regarding Mr. Hardtner you can arrange to endorse the note, or pay the $2,500 but I understood that the Railroad and Lumber Company went together; however, as he has given me figures pertaining to the cost of getting the logs to the saw mill via the Railroad, it may be sufficient that I get the interest in the saw mill alone. He cannot then consistently turn the profits over to the Railroad as against the saw mill. To a certain extent, you can certainly use your best judgment in the matter.

"I notice you now have the steamer 'You & I' in the service of the Railroad Company and secured a good load for her first trip, all of which is satisfactory, and I trust you have the papers in such shape that there can be no question that the steamer has been placed in service, getting and delivering freight between Coushatta and Lake End for and on account of the Railway."

Bill tried to bargain with the Cotton Belt Railroad to get the Silver Lake Lands south of Shreveport.

December 4, 1898

"Hope you are getting lots of freight on the Railroad, and hope the water keeps low. After considering the Silver Lake matter and Cotton Belt interest, then I conclude that my offer to Edson was on liberal lines; and Mr. Harding if he considers my proposition and map must see that it is the easiest solution as it grants them what they want in St. Louis and additional crossing facilities at the Cotton and Commerce Street intersection."

Bill tried to purchase strategic lands between Black and Saline Lakes to keep the Buchanan Railroad from entering the area.

December 9, 1898

The First Avenue Hotel
High Holborn, London

Gents:

"Situation peculiar. Mr. Easton at Biarritz (a spa built by Napoleon III for Empress Eugenie), a 2 to 3 day journey from here. Had several interviews wth his Solicitors Boxall and Boxall and offered 2 1/2 per acre cash if necessary but preferably my note for 3 to 4 years at 4% with collaterals. Boxall believed nothing less than $3.00 per acre would possibly tempt Easton to sell. Finally offered $3.00 per acre on condition that he increase the amount by adding cash to make it $150,000 and take my 3 notes for $50,000 each at 4% - 4.5 and 6 years; put this proposition in writing, and Mr. Boxall was to mail and request telegraphic answer. Before leaving office a telegram came saying Mr. Easton's father had just died! If that changes situation or not can't tell. Mr. Easton may now come to London or may not - certainly will have to wait some days. Meanwhile, will go to Germany where I am to get telegraphic answer from Boxall as to movements and wishes of Easton. Apparently no one has made any effort to buy the Lands except some local people in Natchitoches asking Agency to manage Sale and talking of good prices obtainable. Mail etc. will be forwarded to me in Germany, and I will probably be back here in say 8 -10 days.

"Hope matters progress well with you. If you have not done so yet, use the Railroad earnings and not draw from Illinois Telegraph and Savings. *Do not* deposit the Railroad earnings in *one* bank, thereby disclosing the earnings. Do so deposit and draw that *no one* can tell or even guess at the Railroad earnings.

"Try to get most of your lands at or near the strategically important points. Allow no talk or print about our business or intentions refer to me. If you have secured considerable lands already, you may defer Southward till my return. I yet believe I shall get the Easton lands though not sure. Great inquiries at Office Illinois Steel here for Rails."

Good bye,

Wm. Edenborn

Colonel Easton bargains.

December 22, 1898

The First Avenue Hotel
High Holborn, London

Gentlemen:

"Have had a very vexatious day of it; yesterday I had Boxall to say Colonel Easton would take $2.85 per acre if I would take all his lands under quit claim. I had him almost to agree that he would recommend to take 2.50, but this morning he got a letter from one Porter of Natchitoches saying that much inquiry was made for Timberlands and that a good price might now be realized and asking that the Sale be placed with them.

"Few Steamers are sailing as you may get several letters at once. Mrs. Edenborn is in Atlanta. If I close this week, will return there. Hope you are all well.

Yours,

Wm. Edenborn

London Agency
American Steel & Wire Co.
Billiter Square Buildings, Billiter Square
December 21, 1898

Gents:

"Have received several letters from you and no doubt will find a number on my return to New York. I am sorry to say I have not yet seen Mr. Easton! When I was here before he was in Biarritz, and while talking to his Solicitors, word came of Easton's father's death; so the matter had to be postponed. I went with Mrs. Edenborn to Westphalia and have examined some immense improvements made by the Government, and also a patent appliance to a Nail Machine whereby a round wire fed into the Machine can be made to produce a square or oval Nail, which I consider of much moment to the Wire Company inasmuch as such Nails are wanted by the Export Trade."

Yours,

Wm. Edenborn

The promoter of the Louisiana & Northwestern Railroad told Bill, but not in writing that if he would build a line through Black Lake Narrow to Chestnut, named after a Buchanan executive, he would connect with Bill's Railroad at Chestnut. Bill built a line to Chestnut, but the Louisiana & Northwestern promoter decided not to connect at Chestnut and instead crossed Bill's tracks at Hagen. Bill's line to Chestnut is now Louisiana Highway #9.

Montgomery, La., was located on the Red River near Bill's railroad station which was on a nearby hill. The town was slowly being washed away by the Red River so the town was moved up near Bill's station, named after the land seller, Judge Frank Machen.

By October, 1901, the line to Pineville was completed and placed into operation, complete with passenger and freight service to Shreveport. Because the bridge was constructed to accommodate both railroad and automobile traffic, Pineville grew rapidly and became a suburb of Alexandria, the crossroads of Louisiana. A plaque in Pineville attributes the city's growth to the Edenborn line. The Pineville bridge was completed in May, 1902.

It was during this time that Bill and Sarah entertained a guest, Juliette Doerr, from St. Louis, and introduced her to Bill's partner in the Urania Lumber Company, Henry Hardtner. Soon after they met, Juliette and Henry married, to begin the Hardtner dynasty in lumber in Louisiana.

South to New Orleans

When the road to Alexandria had been completed, Bill and Sarah enjoyed their home there. One evening the Mayor of New Orleans came to call. Paul De Capdeville had enlisted in Boone's artillery during the war, had been wounded and captured by the Federal Army. His captors had paroled him to his home in the Pontalba Apartments, certain that he would not attempt to return to battle, because a parolee, if caught, would be shot. However, De Capdeville had escaped through the swamps and the Federal lines to rejoin the Confederate Army in Mississippi.

Now, the Mayor told Bill, the area was ready to rebuild; New Orleans needed greater access to the Red River and to the wheat of Kansas. He promised Bill access to the wharves on the riverside in front of Audubon Park; and he cautioned Bill that the promoter railroads already there would attempt to block Bill's access.

Audubon Park was the site of the Etienne De Bore' plantation; Bore' had invented a process for granulating green sugar cane because the Louisiana growing season was too short for the natural maturation of the cane and the cane had to be harvested before time. Eventually Bill's railroad would ship more cane than any other line. Bill tried, too, but in vain, to improve Bore's method of granulating sugar.

Clarence Ellerbe surveyed the land from maps and horseback. He considered it to be practicable to continue to Baton Rouge (Red Stick) along the western side of the river and to ferry the trains across the river at Baton Rouge.

The Texas Pacific controlled the land ahead of them, so rather than confront Gould, Bill decided to cross the river to Angola where they could haul the produce of the giant prison sugar plantation and plant.

The transfer across the Mississippi was extremely hazardous for several reasons: first, the shoals shifted there, causing boats to sink; flood water could wash out the track aprons on shore which connected the transfer boats; floods could also cause crevasses in the levee in the connecting ports; and fogs would delay transfers. This transfer was so hazardous that no insurance company would insure the boats or the cargo between the connecting points. Bill was determined that this railroad would go through, so he insured the transfer boats and cargos himself.

And yet the transfer boats were far less expensive than building a bridge of the Atchafalaya or the Mississippi. These bridges involved the same deep river caisson work that James Eads had described in St. Louis. The railroad had to penetrate the Tunica Hills which were of volcanic ash; and they had to conquer the bottomless morass south of Baton Rouge. Bill was determined to overcome these and other obstacles; while Clarence instituted the surveys and hired personnel, Bill sold steel stock selectively to raise capital. And even while the work progressed, Bill knew of the threats of hurricanes and worried about the tactics of the promoters who waited ahead of him.

The St. Louis and San Francisco saw what Bill was attempting. To forestall him, they claimed the same wharf frontage for themselves at New Orleans; as a safeguard, Bill purchased several more strips of land from the Mississippi River to Lake Ponchartrain to assure access from the river to the lake. These arpents of land (present site of Edenborn and Hessmer Avenues) had previously belonged to the Francois de LaBarre plantation, mentioned earlier.

By September, 1901, Bill and Sarah had completed the line to Mansura, Louisiana. In the fall of 1902, construction of the Winnfield-Aloha branch was begun; that same fall, surveys were made for the line from Angola to New Orleans.

In 1903, Randolph, the Shreveport attorney, incorporated the Louisiana Railway and Navigation Company to purchase the assets of the Shreveport and Red River Valley Railway, and to continue the construction of the line to New Orleans.

It took a year to complete the 20 mile line from Aloha to Winnfield (the home of Huey Pierce Long). That same year, 1903, saw the completion of the cypress line to the "town on stilts," Naples. Bill had named Naples after the Neapolitan boat song, "Santa Lucia," because there was a long stretch of water before Naples: "Now, 'neath the silver moon, ocean is blowing."

During this time the Port Hudson-Baton Rouge line was put into operation. Inclines, or aprons were built at Naples and Angola to connect the tracks for the long journey across the Mississippi. Whenever the railroad was extended to a town of any size there was always a ceremony to celebrate the event. Bill and Clarence always presided over these ceremonies, and, although they were only token drinkers, they enjoyed the entertainment with their wives.

Then levees had to be built to shelter Sarto, a very important shipping point during the cotton season, which lay adjacent to the Mississippi River. Much of this section of line was constructed of huge cypress pilings, measuring 75 feet long, which never touched bottom, but held up the tracks by a friction force effect. These cypress pilings were among the first "friction pilings" used in wet alluvium, such as is found in south Louisiana; these cypress pilings were used to support buildings in New Orleans. Hand hewn by the old time professionals who used the broad axe, the cypress pilings cost Bill $1.50 per tree; Bill found that driving a piling through a swamp was less expensive than grading and maintaining an earthern right of way that constantly sank into the swamp.

When the line from Sarto to Alexandria was completed Bill Became acquainted with Firing Engine 68 and its engineer, Johnny Williams.

At the same time that the line to Naples was being built, Bill supervised the construction of a line south from Angola, the site of the penitentiary, eight miles across the Mississippi to Baton Rouge. For a time, this line operated independently; it went through the old Barrow Plantation were Bill had met the famous Dr. Sol Smith 30 years earlier. Here, Bill laid out the town of Bingen, whose namesake was the Ruhr Valley hamlet. However, Bingen, like some of Bill's other towns, reverted back to the forest and farmland, glimpsed only fleetingly by the passengers of the Louisiana Railway. Telegraph wires were laid in submarine cable along this line between Naples and Angola.

During this time the northward shipment of freight and passengers came over Morgan's Louisiana and Texas Railroads and up to Bill's railroad. Morgan's railroad has since become a subsidiary of Huntington's Southern Pacific.

The water transfer of trains continued to cause problems. The boats would get stuck on the constantly shifting mud and remain there, laden with cargo, for hours until extricated.

One section of the river was especially problematic. This section was called Old River and had been formed by the Mississippi scouring out a channel through a farm called Torras' Furrow in 1860.

The channel had cut Turnbull Island off from the mainland. During the times when the water rose, the bears and deer would swim away from the island, but the rabbits remained. Bill and his hunting guests often shot 100 rabbits in a day. Later, Theodore Roosevelt was to use Turnbull Island as a holiday camp.

TEN

South of Baton Rouge

As they continued south of Baton Rouge, Bill and Clarence had to cross several large swamps. The first of these was the McElroy Swamp which was 15 miles long and apparently bottomless, for anything heavy sank, and even the longest pilings did not touch bottom. In order to get fill for the ballast and right of way, a novel engineering approach was conceived. Large cypress trees, as much as 15 feet in diameter and 2500 years old, were felled and used as corduroy. On top of the floating logs, which were virtually impermeable to rot, five to six feet of dirt and sand were topped with iron ore gravel from Bill's pits (located at Bruder, north of Baton Rouge). Then several feet of pea gravel were applied. These hard cypress ties, except for mechanical wear and tear, would last almost forever.

To build the floating railroad through the McElroy Swamp, Bill and Sarah laid out the towns of Nettie and Edenborn near the rivertown, Gonzales. The workers who built through the swamps lived in these small villages; the swampers who felled and floated the giant logs, lived nearby. The giant cypress logs were also used as pontoons in the bridge building.

It was a long and arduous task for the technicians who built this floating railroad; the bayous were infested with all kinds of insects, snakes and alligators. The technicians had to keep their bodies completely covered to protect themselves against one kind of mosquito that bit during the day and another that bit during the night. Yet, despite these difficulties, this section of line became the best riding track on the railroad.

The arrival of the railroad to the area guaranteed year round employment to the cane harvesters at the Burnside plantation and to the local population. The iron bands of Bill's railroad heralded the end of the old steamboat days on the Louisiana rivers and bayous. Years later the knockabouts in the terminal at Edenborn, Louisiana, would reminisce about the ghosts of those old steamboats: "one-arm John" with one chimney; "the dirty bell" which belched a black cloud of smoke; and how they could identify any of them during the darkest of nights by the sounds of their whistles. From now on excursions to Baton Rouge would be on the special trains that the iron bands carried.

The ride through McElroy Swamp took almost an hour; the trees so thick that during the daylight the electric lamps had to be switched on; the mosquitos thick and fierce that bare arms had to be wrapped in newspaper to protect the skin. Bill and Sarah, in an attempt to beautify the swamp by creating a facimile of Shaw's Gardens, planted water lilies along the right of way. Unfortunately, the lilies grew so profusely that they soon clogged the slow moving waters and blocked the railway itself and had to be removed by men in piroques at great expense to Bill.

That area was known as the German coast and had been settled by immigrants who adopted French sounding names: Milton Berger was changed to Mil De Bergue; Gottchaux was another derivative name. Bill named towns after people he'd met; Barman was named after an alderman from Gonzalez, Louisiana; and Bill named towns after the various land vendors as well as after historic figures such as Godfrey de Bullion, who'd led the Second Crusade; and, of course, there was Edenborn, Louisiana.

Then the Frisco Railroad attempted to restrain Bill from getting passage and wharfage on the New Orleans riverfront. Paul De Capdeville instituted a lawsuit against the Frisco Railroad on Bill's behalf; this lawsuit was completed some time later by Martin Behrmann, the mayor of New Orleans and at one time the ward boss of the Choctaw Club, the democratic political club of New Orleans. When the suit was won, the Edenborn line went down to the river and the city built several large grain elevators to hold the grain that had come down from Kansas to be processed and transshipped.

It was during this time that Bill got an idea that didn't work. Sarah and Bill were dining one evening in the French Quarter near the Lafitte Bar and they asked an organ grinder to play "Sorrento," a song that Bill particularly liked and had even named a town after. Because World War I was in the news and because labor was at a premium, Bill and the organ grinder, V. Bommarito, discussed the possibilities of training monkeys to perform certain, unskilled kinds of labor. The organ grinder and three monkeys accompanied Bill to his research farm in Emden, Louisiana, where they trained the monkeys to pick cotton. That part of the experiment was extremely successful. The part that failed was this: the monkeys insisted on shredding the cotton away from the seed so that they could eat the seed. No amount of training could stop the monkeys from shredding the cotton, so Bill had to give up his idea of cheap monkey labor.

Once again Bill was to encounter the memory of Jay Gould in the railroad business. Bill had bought a Katy line that ran from Shreveport to Dallas; however, it had been badly rundown and poorly maintained by the government. And to add to the problems, Jay Gould who controlled the Missouri-Pacific Railroad, also maintained control of the Katy line. And when there was freight to be shipped, he shipped it cheaper on the Missouri-Pacific line, causing the Katy line to lose money.

New Orleans Jazz

Bill and Sarah often took a hack ride to Storeyville, an area of 38 square blocks named after Congressman Storeyback, during the Gay Nineties. This area had been set aside for the "working girls" of the city. The "ladies of the night" were prohibited from crossing the boundaries of this area.

Storeyville was the center of jazz and ragtime music. Bill and Sarah had learned to enjoy that music in St. Louis, at the Rosebud at 2220 Market Street, which was operated by Tom Turpin, the famous pianist-composer who had written the "Harlem Rag". Another musician at the Rosebud had been Scott Joplin who had composed the ragtime opera, "Tremonisha" which Bill recognized as a masterpiece. Bill and Sarah had learned how to play the piano, and Bill enjoyed playing "Turkey in the Straw".

Another musician that Bill enjoyed was Kid Ory, who often performed at Anderson's, the place that Bill and Sarah frequented in Storeyville. Kid Ory performed his song, "The Muskrat Ramble" with his side men, Johnny Dodds, King Oliver, and Laten Armstrong, and would tease Bill about a railroad line named The Louisiana Railway and Aggravation Company, because of a train that was always late.

Other musicians at the Anderson restaurant were Boyd Senter and Buddy Bolden, who played a horn too loudly for Bill's taste.

ELEVEN

The Importance of Bill's Railroads to the Towns They Served

These articles from the COLFAX CHRONICLE of 1903 tell of the effects that the Edenborn railways and lumber activities had on the Colfax, the seat of Grant Parish, Louisiana.

"Better Freight Service on the Valley"

"A switch engine is now located regularly in the Valley railway yards here so that cars with perishable goods can be more promptly switched and put in position to be unloaded. The engine and crew are in charge of Conductor Flanders, who will move his family here from Texas at an early date.

"This switch engine also makes regular trips to Mansura with a mixed train, freight and passengers, twice a week — Tuesdays and Fridays — and oftener if business justifies. The hour of leaving here for Mansura on the days mentioned is 11 a.m., and returning at 4 p.m.

"Another change on the Valley railroad between here and Shreveport is, that the mixed freight and passenger train formerly came as far as Colfax, now comes to Alexandria. It leaves Shreveport at 6:30 a.m., and arrives at Alexandria at 6 p.m. Leaves Alexandria at 4 a.m. and arrives at Shreveport at 5 p.m. This is a daily train except Sunday. it will be a great convenience to merchants as perishable goods will come on through instead of being laid over at Colfax several hours as heretofore.

"The local train that formerly left here daily at 7 p.m., going as far as Colfax, has been discontinued. — Daily Town Talk."

"Reduced Rates for Baseball Games"

"Tickets at reduced rates will be on sale for regular passenger train into Shreveport on above date. Special train will leave Shreveport at 7 p.m. and run to Campti. Tickets sold at and north of Campti, Louisiana, limited for return passage on date of sale. Tickets from points south of Campti will be limited for return passage until Monday, July 27th. See your ticket agent for any additional information. — C.E."

"New Orleans and the Louisiana Railway and Navigation Co."

"We notice in the New Orleans papers that the Louisiana Railway & Navigation Company is asking that city to grant it the necessary rights of way and franchises to enable to enter and do business in that city.

"It would seem at this distance that the grant of such right, properly guarded would be gladly made by the "future great" city. The above named railroad was first organized as a local road — the Shreveport and Red River Valley Railway Company

— to run from Shreveport to Alexandria. But since the great movement of all the great East and West systems to get into New Orleans, it has been seen that the old Shreveport and Red River Valley line can be turned to good account to enable several great systems to reach that city. Hence the recent reorganization, enlargement, increase of capital and change of name to "Louisiana Railway & Navigation Company".

"It is well known that this change of name and large increase in capital stock has great significance. It means that this railway line will at once be pushed into New Orleans; that as soon as constructed into that city the traffic of at least two other great railway systems, embracing 10,000 to 15,000 miles of road, will be run over that line into New Orleans; that there is the best of reasons for believing that two or three other great systems will also come into the same arrangement and avail of the same means to reach the great Gulf Port; and, finally, that it is in contemplation to double track the Louisiana Railway & Navigation Co's. road, for which purpose right of way has already been secured from Angola to Baton Rouge.

"The line of this road is located from Baton Rouge to New Orleans, and track laying is in progress above Baton Rouge. All the cities, towns and country people along the lines of this popular and much needed railroad, and all of North and Northwestern Louisiana, are vitally interested to seeing this railroad, which gives them the shortest and most direct communication with the capital of the State and with our greatest commercial seaport, secure entrance into the Crescent City. They are watching the action of the New Orleans city council on the ordinance granting the necessary rights of way and franchises to this (our own) railway with great concern and interest.

"The cities, towns and country along the line from Shreveport to New Orleans have well performed their part toward getting this important railway constructed and we all sincerely trust our great city will be as generous and will do her part without stint or hesitation. It not only means 10,000 to 30,000 more miles of railways for New Orleans but it means a new steamship line to run from Western Europe to that great and growing city, in direct connection with these great railway systems."

"Chain Gang from Angola"

"A squad of 15 convicts passed through Colfax last Sunday morning on the Valley train on their way to Taylortown to work on the levee. They were in charge of Sergeants Robertson and Forquer."

Bill formed the Colfax Lumber & Creosote Company to creosote railroad ties; Colfax & Creosote is a major company today. He named a town after J.H. McNeeley (mentioned in the following article) who was president of the Bank of Colfax at that time.

"Big Sawmill for Colfax"

Wednesday afternoon Mr. Clarence Ellerbe, the energetic secretary and treasurer of the Louisiana Railway & Navigation Company, arrived in Colfax by hand car. He had just completed an inspection of the company's interests at Verda, Atlanta, and other points along the Colfax & Northern railroad, most commonly known as the Winnfield branch.

"Mr. Ellerbe announced that Mr. Edenborn and the Colfax Lumber Co. had closed a trade with the Big Pine Lumber Co., by which that company gets possession of a large body of timber lands. The Big Pine Lumber Company is a substantial concern composed of G.W. Fouke, Hugh Corry, Ben Collins and other citizens of Texarkana, Ark., who are reputed to have $300,000 capital behind them. They have been very successful in the lumber business, having already had several sawmills in operation.

"This company proposes to erect a mill at Colfax of sufficient capacity to cut a million and a half feet of lumber every month in the year. Mr. Ellerbe announces that they have put up a cash deposit with this company of $15,000 to be forfeited if they do not have their plant in running operation by the first of September next. Mr. Ellerbe further announced that the location of the mill had not been determined, but that it was to be at some point on the road from Colfax to Verda, yet to be selected. He said Mr. Edenborn stood ready to give them 160 acres for a mill site at almost any point where he owned the land, but the parties desired to locate at Colfax, provided they could get land sufficient for their purpose, and that they wanted not less than 60 acres.

"In view of the importance of the proposition, Mr. Ellerbe asked that a mass meeting of the citizens of Colfax be held in the court house at 5 o'clock that afternoon to consider the offer.

"Pursuant to this suggestion the citizens assembled in the court house at 5 o'clock. Capt. C. H. Teal was called upon to act as chairman of the meeting with H.G. Goodwyn and W. B. Clarke as secretaries.

Capt. Teal stated the object of the meeting, saying that he was willing to sell sixty acres of land to the company at the very reasonable figure of $50 per acre and that he would donate $500 of the $3,000 purchase price of same. The land in question lies east of the railroad track, just north of the brick yard and has been approved as an acceptable location for the mill.

"Mr. Ellerbe being present was called upon by the chairman to state the details of the proposition more fully. He detailed the particulars already stated, stating that he had in his pocket the contract and agreement of forfeiture referred to, and he thought the proposition was one the people of Colfax would do well to accept. He and his company felt great interest in the progress and prosperity of the town because its prosperity meant business for their railroad, but aside from the financial aspect, they had a true and kindly interest for our citizens, who had always treated them handsomely. He had especially worked up the present enterprise in the interest of the town, and he sincerely hoped the citizens of Colfax would make sure of its location here, because its prompt acceptance would settle the matter at once and establish confidence in the stability and future of the town.

"J.H. NcNeeley, Dr. W.J. Roberts, J.W. Duncan, H.G. Goodwyn, and others expressed their views of the proposition. In the course of the discussion it was announced that Mr. Hugh Corry, one of the principal stockholders in the Big Pine Lumber Company, was at the hotel.

"H.G. Goodwyn moved that Mr. Corry be invited to take part in the meeting. J.H. McNeely moved as a substitute that the chair appoint a committee to invite him to be present. The substitute being accepted, the chair appointed J.H. McNeely, J.W. Duncan and R.R. Jeter as said committee.

"Mr. Corry being introduced by the chairman, thanked the citizens for the invitation to be present, and said that if the proposition they had under discussion was accepted he intended to bring his family here and be in our midst. He is a Democrat and a Southern man, and not stranger to our institutions. The mill his company expected to build would give employment to two hundred men, and its location here would aid materially toward the support of our institutions.

"On motion of J.H. McNeely, duly seconded, it was revealed that it is the sentiment of the meeting that the citizens of Colfax should do all they can to assist and encourage the Big Pine Lumber Company to locate their mills here, and that in furtherance of this end the chair is requested to appoint a committee to solicit subscriptions to purchase the site desired.

"The chair appointed Earl Roberts, J.W. Duncan, Dr. W.J. Roberts and R.R. Jeter as the committee to solicit subscriptions for this purpose.

"The meeting then adjourned."

"Another headline read: "Big Sawmill at Colfax Is Now a Certainty". This article mentioned that the president of the Colfax Bank, J.H. McNeely had had a town named after him. And: "Furthermore, Misters G.W. Fouke, president, and Hugh Corry, secretary of the Big Pine Lumber Co., have been at work with a force of men this week getting things in shape to rush the building of the big saw mill which has been located here. Also the engineer of the Valley railroad came to Colfax on Wednesday and laid off the switch and sidings for the use of the mill."

"Valley Road Starts Freight Service in New Orleans"

J.A. Saxton, General Agent of the Louisiana Railway & Navigation Company in New Orleans, has announced that his Company is now running four freight cars a day out of New Orleans over the Southern Pacific as far as Alexandria. At Alexandria the cars are sent respectively over the Arkansas Southern and Louisiana and Arkansas, by way of Winnfield, and north and south over the Louisiana Railway & Navigation Company's lines. One car goes over the company's lines to Shreveport, another goes south over the line to Sarto, (Big Bend) on the Mississippi River, to which point the company's line is completed. From Alexandria two cars go north to Winnfield, from which point one goes north over the Arkansas Southern, and the other north over the Louisiana and Arkansas to Hope.

"The freight which is handled in this manner goes out from New Orleans under through rates, and it goes into a section which was heretofore been very difficult of access to New Orleans merchants. The best route heretofore was by way of the Yazoo and Mississippi Valley to Vicksburg, and thence over the Vicksburg. Shreveport and Pacific west towards Shreveport, connecting on that line with the Louisiana and Arkansas and the Arkansas Southern. By the Louisiana Railway & Navigation Company's route the freight goes directly into the territory, saving a distance of nearly 200 miles. This territory, which is thus upended up to New Orleans merchants, is both valuable and extensive. The Louisiana and Arkansas, extending from Hope, Ark., to Jena, is 187 miles long, and all of the territory which it traverses is opened up by direct connection with New Orleans. Points in the Arkansas Southern between Eldorado and Winnfield, a distance of 101 miles, are also opened up. Points on the Louisiana Railway & Navigation Company's line from Sarto to Shreveport, a distance of sixty-five miles, another valuable territory is opened up.

"On the Louisiana and Arkansas such points as Springhill, Scott Valley, Minden, Sibley, Jamestown, and Goldonna, in Louisiana, and Stamps, Hope and Taylor in Arkansas, are opened up. On the Arkansas Southern, such points as Camden and Junction City, in Arkansas, and Bernice, Dubach, Ruston, Quitman, Jonesboro and Wenona are reached directly from New Orleans.

"This territory, it is said, has been closed to New Orleans merchants on account of the fact that the roundabout way freights had to go made St. Louis, Kansas City and Chicago, in effect, nearer markets. Since the establishment of the through rates an effort will be made to secure for New Orleans all of the business of this territory which is very rich, and which offers great advantages to New Orleans merchants of every class.

"Mr. Saxton also announces that through rates have been made with the Cotton Belt to all points in Texas, Louisiana and Arkansas and to all points on the Sibley, Lake Bistineau and Southern Railway, which runs down from Sibley to Wilson, in the Long-Bell Lumber Company's territory. These rates were effective on April 9, and identical with the rates over the Texas and Pacific, the Southern Pacific, and the Vicksburg, Shreveport and Pacific. They include such points as Dallas, Corsicana, Sherman and Greenville.

"Mr. Saxton stated yesterday that the Louisiana Railway & Navigation Company had completed its line from Shreveport to Sarto, which is just opposite Angola. Construction work is going on between Angola and Bayou Sara, and on this side of Bayou Sara the bridge over Thompson's creek is being worked on. All the necessary steel is on the ground, and the bridge will be completed in about five weeks. When these gaps are connected there will remain only the line from Baton Rouge to New Orleans uncompleted."

"The Big Pine Lumber Mill"

"All is activity and bustle on the grounds of the Big Pine Lumber Co. The carpenters have the big timbers of the first and second stories of the immense mill all raised and in place, and are getting ready to put on the roof. Three or four car loads of engines and machinery for the mill have been received, and some of it is being unloaded.

"Mr. G.W. Fouke, president of the company, and other stockholders, have been in Colfax this week, attending to the drafting of a new charter. The company has been re-chartered under the laws of Louisiana, and the new charter will be found among the advertisements in our columns this week. The company is organized with a capital of $150,000 with $127,300 subscribed and paid in at this time. The various stockholders and amounts held by each can be seen by reference to the charter.

"About one hundred men are now at work, in the woods, on the mill and about the grounds. Mr. Hugh Corry, secretary and general manager, with Mr. Harry Fouke as clerk and assistant, has supervision of everything, and all departments move with the regularity of clockwork. Mr. Merrill Sheppard is superintendent in charge of the erection of the plant, with Misters Tom Duke and B.J. Ross as his assistants. Mr. Joe Fusswell, for over twenty years woods foreman with Mr. George Fouke and his associates, has charge of the woods. The well kept appearance of his nearly one hundred oxen denotes his ability to keep up his end of the line."

"Charter of Big Pine Lumber Co., Limited"

"Be it remembered, that on this 10th day of May, 1901, before me, John A. Williams, Notary Public in the parish of Grant, State of Louisiana, personally came and appeared the persons whose names are hereunto subscribed, all of full age, who declared that availing themselves of the provisions of this State, relative to the organization of private corporations and especially the provisions of Act No 36, of the General Assembly of Louisiana of 1888, approved June 29th, A.D. 1888, they have covenanted and agreed to form and organize themselves and such other persons as may be hereafter associated with them into a limited private corporation with the rights, privileges, and under the name and title, and under the stipulations, agreements and conditions for the uses, objects, and purposes hereinafter set forth in this act of incorporation.

"The Big Pine Sawmill at Colfax is to be one of the biggest and most-up-to-date mills in this region. It will run two main saws, besides innumerable small saws. The main building is completed, the boilers are in place, and the machinery is being rapidly assembled, the intention being to commense sawing in about four weeks. Nothing has been done on the planing mills as yet, but we understand operations on it will be begun in a very short while."

"A 'Philanthropic' Town"

The Big Pine Lumber Co.'s extensive operations in Colfax will begin in about ten days. Their locomotive and above thirty logging cars are at Colfax, ready for business. The locomotive has "Colfax" painted on it. The commissary, twelve or fifteen houses for the hands to live in, and a number of other buildings have been completed. Two deep wells affording an abundant supply of water, and a mammoth underground cistern 20 x 30 by six feet deep which is to be cemented so as to catch and hold rain water for drinking purposes, is also nearing completion. A canal or ditch, 400 feet long, ten feet wide and eight feet in depth, has been cased inside with two-inch lumber, is half full of water, and ready for the saw logs to be dumped into it. The three smoke stacks are up, the big mill is done, with all its six boilers, huge engines, machinery, etc., in place and the engineers calculate to "turn her over" during the ensuing week in order to test everything and see that it is in good working order. Also the foundation for the Planing mill has been put down, a good portion of its machinery is here, and the building is to be rushed to completion at once.

On last Sunday, Mr. G.W. Fouke, president of the company, together with Mr. Hugh Corry, its hustling secretary and general manager, and Mr. Harry P. Fouke, paymaster and general factotum, went to Shreveport on business. The last named is to visit the World's Fair at St. Louis and may also get to New York and other northern cities, taking about a month's vacation before returning."

The SHREVEPORT TIMES wrote:

"Among the prominent people in the city yesterday were Co. Geroge W. Fouke, Judge Hugh Corry, and Harry Fouke, of the Big Pine Lumber Co., the mammoth new plant just put down at Colfax. Col. Fouke, who is a Texarkana millionaire capitalist is president of the concern, and Judge Corry, a wealthy lumberman, formerly of Arkansas, as vice president and general manager, while Mr. Fouke, the younger, is general utility man.

"The trio spent the day here and left for Texarkana in the evening, where they go on a business trip.

"The big plant at Colfax is finished and will open up for the first day of August for a twenty-year run, most of the time running day and night. Judge Corry says the reason they will run so constantly is not for the purpose of making money solely, but that they are anxious to furnish the world with all the high grade Louisiana yellow pine possible before their usefulness expire. In addition to being a lumberman and a politician, Judge Corry is also, as may be seen, a big hearted philanthropist."

The NEW ORLEANS PICAYUNE of December 3 had the following to say about the Valley railroad:

"Trafficmen generally admit that the Red River Valley, now the Louisiana Railway & Navigation Co., is little short of a wonder in getting new business. By its Alexandria connections with the Southern Pacific and the Texas & Pacific, it is hauling freight in and out of New Orleans at an astonishing rate.

"In line with these new and improved traffic connections, a conference was held in the city yesterday between the trafficmen of the Edenborn lines and those of the Southern Pacific that will still further foster the handling of freights up through this new country pierced by the Red River Line.

"E.D. Marshall, general freight agent, and H.B. Helm, general superintendent, were in the city. They conferred during the day with General Freight Agent, Charles S. Fay of the Southern Pacific and other connecting lines here.

"The Red River expects to have its own line into the city within a year," Fay said. "Until that time, it proposes to keep up the close relations with the connections at Alexandria."

"The Valley to Build a Road from Alexandria to Jena"

Some two weeks ago a railroad surveying corps started at the Red River bridge of the Louisiana Railway & Navigation Company, on the Pineville side of the river. The Town Talk tried to learn what was intended by the survey, but without success. We are now reliably informed that the corps is surveying a line of road for the Louisiana Railway & Navigation Company from the Pineville end of the Red River bridge of the Company from here to Jena, Catahoula parish, a distance of 28 or 30 miles. The survey has already been nearly completed, and contracts for building the new line will be set as soon as possible and pushed to completion. Major Poindexter Dunn is in the city, and will be busily engaged for the next several weeks in securing rights of way for the road from Alexandria to Jena. This new road will open up a fine territory for Alexandria and will bring much more business here for our wholesale as well as our retail merchants. Jena is the present eastern terminus of the Louisiana & Arkansas Railway, which is now building into Alexandria from a point 9 miles east of Winnfield. We judge that the new road will pass three or four miles west of Holloway Prairie.

Major Dunn is the promoter and solicitor for the Alexandria Bayou Macon & Greenville Railroad, which is proposed to be constructed between this city and Greenville, Mississippi, via Harrisburg, up the Macon Ridge and to a point on the Mississippi River opposite Greenville. The distance is 180 miles. Active work on the construction of this road is to be commenced in the early fall, at the Mississippi River, near Greenville, and pushed in the direction of this city.

"At Greenville the proposed road gives connection with the Rock Island, the Missouri Pacific, the Illinois Central, and the Southern System, and it can readily be seen of what great value this new road will be to Alexandria. The Southern System is to put in the transfer at Greenville.

"The A.B.M. & G. will cross the V.B. & P.R.R. at Delhi, Louisiana, which makes that town a junction of importance. An election will be held there on the 23d of May which will give a 5 mill tax to the road for ten years."

"The Valley Railroad Will Go to New Orleans"

"This has been a busy week for the Shreveport and Red River Valley railroad people, getting matters in shape for their ocean export trade. Vice-president McElvried, Gen. Supt. Ellerbe, and the attorneys for the road have been in New Orleans all this week looking after terminal facilities, and Mr. Edenborn is expected to be there early next week to confirm and approve arrangements looking to that desirable end.

"The people of this valley will be pleased to know that they are to have direct connection with the metropolis of the South, and that our railroad is to be one of the great trunk lines connecting New Orleans with the North and West, thus linking us with the great busy world of trade and commerce. The PICAYUNE of January 28th, says, editorially:

"The City Council did not last night, as it had been given out would be done, hand over the public streets, a public place which is really a park plot, and the city's Belt Railroad, to a private railroad company. There was a hitch in the arrangements that prevented the consumation of an outrage upon the people of the city, and the matter was postponed. The pretext for this proposed making away with public property in this prodigal manner was that the city had invited railway companies to make their terminals here, and but one company had responded, and it was due this company to give it all that might have been divided out among others.

" 'Fortunately, a postponement was had of the contemplated proceedings at the meeting of the council last night and then an interesting and important surprise was sprung upon the city solons. This came in the shape of a formal proposition of the Shreveport and Red River Valley Railroad, asking for accomodations and facilities in the city for the proper transaction of its business. The Company was represented by its officials and by counsel, and its petition sets forth," among other matters, that the Shreveport and Red River Valley Company is operating a railroad from the city of Shreveport to the town of Mansura, in the parish of Avoyeures and is desirous of extending the same to the city of New Orleans. That, with said object in view, your petitioner, at great expense, has acquired property on the east bank of the Mississippi River within the city limits to be used for wharves and elevators, and to furnish facilities for exporting the products to be brought to New Orleans over its lines.

" 'That petitioner has purchased a large area of lands suitably located within the city limits to be used for depots, warehouses, roundhouses and general railway purposes.

" 'That petitioned land made said investments under the belief that it would be allowed to use the best tracks of the city of New Orleans on terms as favorable as would be accorded to any other railroad company.

" 'That an ordinance is now pending before the honorable City Council, giving certain rights and privileges to the New Orleans and San Francisco Railroad Company and especially with reference to the said belt tracks; that petitioner, as an owner of property and as a railroad company seeking to enter the city of New Orleans, desires to be heard in regard to the details of said ordinances, and particularly as to said belt tracks. Therefore, petitioner prays that said ordinances be recommitted, or that such other appropriate action be taken as shall allow your petitioner to be heard on the matter involved therein."

"This company does not come heralded with the announcement that it has unlimited cash which it is ready to hand out wherever it would do the most good to its own advancement, but there is no doubt that it is the advance guard and representative of some great trunk line system that desires to make its ocean terminal in New Orleans. It states, as does the "Frisco", that it has purchased extensive property in the city, and that it has secured space on the riverfront, so there is no lack of solidity in its claim for attention, and this it should have at the hands of this city."

The SHREVEPORT TIMES of April 25th asked: "Has Frisco Railroad Got Control of Valley?"

"Some time ago disposition of both the Kansas City Southern and the Red River Valley Railroads was a mooted question. The TIMES made the first report of the acquisition of the latter road by the Joaquim Frisco interests. That report has subsequently been supported by the arrival of the travelling representative of the passenger division of the Frisco system; he will be in Shreveport for the purpose of establishing acquaintances and connections in the city. The claim is further advanced by the fact that the Frisco system has been given a field in this territory by virtue of the opening of the Thebes bridge by the Cotton Belt lines, and a hard campaign will be made for traffic out of New Orleans via this district. The construction of the new part of the Red River Valley road into New Orleans is expected to have been completed shortly after the first part of the new year."

And then there was the newspaper editorial that said: "We would respectfully suggest to Mr. Edenborn that the name Louisiana Valley Railway & Navigation Company was a better sound than the name Louisiana Railway & Navigation Company, which as been adopted by his new organization. The latter has a sort of bob-tail sound. If any future changes are to be made in the name, we hope our suggestion for a more rhythmic and descriptive one will be considered.

And then one of Bill's employees was murdered:

"Jim Brown, colored, was murdered at Tyrawley, a station on the Valley railroad just across the line within the limits of Rapides Parish, Monday night about 7 o'clock, by Garfield Mills, also colored. The facts are as follows: Some time previous to the murder, Mills had a pistol stolen from him. He accused Brown of the theft. Brown denied it and pulled his pistol out of his pocket and asked Mills if it was his, and Mills said it was not. The affair dropped there, but later Mills went to the camp where Brown was working and again accused him of stealing the pistol, and cursed and abused Brown. Brown told him to stop cursing in the presence of the foreman's wife, who was a white woman. Mills left, vowing vengeance. His chance came Monday night when Brown was on his way to Colfax to see his folks. As the train stopped at the Tyrawley station, Brown got off to assist in unloading some baggage. Mills saw him and asked, "Is that you, Jim?" When Brown answered, Mills began firing; he put five pistol balls into his victim."

WITH ROOSEVELT ON TURNBULL ISLAND

Turnbull Island, which consisted of several hundred square miles of pristine forest, mostly hanging cypress, had originally been a peninsula between the Atchafalaya and the Mississippi River.

The Indians, who believed that the island would always belong to them, said that the Divine had placed the peninsula there to prevent the Mississippi from reverting to its former route along the Atchafalaya River but to continue toward New Orleans.

The peninsula became an island when a farmer, named Torras, plowed a furrow at one end of the peninsula and the rivers rushed through, cutting the island away from the riverbank. The island was named after some settlers from England. These settlers, the Turnbulls, had promoted the first railroad in the United States, which ran from St. Francisville, Louisiana, north about 50 miles to Woodville, Mississippi. The cars and rails were wooden and were drawn by oxen.

Teddy Roosevelt, a guest on the *Sarah*, enjoyed hunting; Bill and he had heard of a great black bear that was exceptionally ferocious. The bear had been named Nate, after the Federal General Nathaniel Banks, who had burned and ravished the area as he had fled from the Confederate Army. Nate, the bear, had killed several hunting dogs, and maimed horses that had tried to ride it down. The local swampers made a serious business of trying to hunt and kill Nate, "the bear that'll eat *you*".

Roosevelt and Bill determined to hunt Nate and to give the skin to the swampers and to mount the head. As they approached the island, they saw the floating remains of the sunken *William Edenborn* and hoped that that was not an omen for their hunt.

They were accompanied by Big Wes, the giant negro chef from the Railroader's Hotel; Big Wes was expert at handling dogs and mules for hunting. He told them stories of how Nate had shaken a cougar "like a rag doll".

Roosevelt, Bill and Big Wes camped that night on the Island. They could only stay one night, for they had to catch the *Sarah* the next evening to return to New Orleans. While sitting around the campfire, they philosophized about man taming nature. Bill admired Roosevelt for building his fitness and stamina and for "wopping the poo" out of the Dakota cowboys who had made fun of his manner of speaking: "Hurry, hasten quickly there" had been the favorite target for jokes.

Roosevelt believed in fair play even for bears, he felt that using hunting dogs gave him an unfair advantage. But time was short, so the dogs had to be used that next day as they set out to hunt Nate.

When they found Nate, he headed away from them and the shrill whistle of the *Sarah* and headed for the Atchafalaya River. Roosevelt held his fire until he could get a fair head or shoulder shot. Then as they approached the river, they could see that Nate was accompanied by two year cubs. Again Roosevelt held fire, this time to avoid hitting either of the cubs. By the time he was able to get a clear shot at Nate the bears had begun to ford the river and it was too late and the distance too great for a "fair shot" at the bear.

When the newspaper reporters heard this story, they coined the phrase, "Teddy Bear" which became synonymous with fair play.

THE WORLD'S FAIR

Bill and Sarah visited the World's Fair; they had contributed to many of the exhibits which pertained to their own industries. First among these interests was the Railroad Exhibit at the Transportation Palace which featured dozens of locomotives from England, France and Germany. There were complete trains in the latest designs, and mechanical engineers from all over the world had come to exchange ideas.

Other exhibits that Sarah and Bill had contributed to were: the coal, cotton and lumber exhibits which displayed the planting, harvesting, and the multitude of uses these products had.

They enjoyed the "city without nails" from the South Sea Islands, built entirely of twine and rope. They met the future Cardinal, John J. Glennon. They met the Veiled Prophet Queen, Stella Wade, whose banker father had the town of Festus, Mo. named after him. They greeted their railroad competitor, Helen Gould, as well as their customers from their music supply company of New Jersey.

Other items of interest were: the present Washington University Administrative buildings were loaned to the Fair for offices; the present Guarantee Electric Company was formed to wire the Fair with wire drawn at the American Steel and Wire plant behind Union Station, the largest terminal in the world.

They met William Weil, the bandmaster of the Fair. But the main attraction was John Philip Sousa who played "Stars and Stripes Forever."

During 1906 while Bill was still an executive of United States Steel, he was visited by his old friend and former attorney, Elbert Gary. Gary was now Chairman of the Board of United States Steel. He told Bill about the model town that had been laid out near Chicago for the steelworkers (the town was named Gary); and of his plans to make the plants as safe as possible for the workers. And Gary told Bill of the welfare plans he had established for the men who worked for the "corporation with a soul." He told Bill about the Slav workers who always cautioned each other, "Tschekai" or "Watch out," how they warned each other to watch out for the cranes carrying beams, or the ladle overhead. Too many men had become part of skyscrapers, bridges or girders. It was Gary's aim to change those conditions.

Arabella Huntington

TWELVE

The William and the Sarah

As we've seen, William Edenborn's business empire grew from the production of barbed wire to steel production, then to the building of railroads and to navigating rivers. In order to avoid lawsuits against the railroad, Bill had his attorney, Randolph, organize the Angola Transfer Company in 1906. The facilities of the company were needed to transfer the freight and passenger business through Old River and across the Mississippi, a record haul of eight miles across water. To accomplish this, the company owned two large stern wheel steamers that attached to transfer barges which carried the passenger cars, freight cars and locomotives.

The first of these steamers was 143 feet long, thirty-three feet, four inches wide, with a gross tonnage of 259 tons, a steel hull, and when loaded drew six feet five inches of water, drawing four feet when empty. This boat towed the barge, *Angola*, built in 1906, measured 256 feet long, 37 feet wide, with a gross tonnage of 647 tons, and could carry 12 freight cars. The *Angola* had a steel hull, divided compartments, and when loaded drew six feet of water, when empty she drew three feet of water.

The first *William Edenborn* had sunk in the fog of Old River; the second was built in 1914. She measured 143 feet long, 33 feet, two inches wide, had gross tonnage of 267 tons; her draft when loaded was six feet five inches, when empty her draft was four feet ten inches. She towed the barge, *Naples*, built in 1908, which measured 196 feet long, 37 feet wide. The *Naples* gross tonnage was 716 tons and she could carry 14 freight cars; when loaded her draft was six feet, six inches, when empty, two feet.

These boats worked twenty-four hours a day when business was available; each had a captain, several pilots and engineers, firemen and deckhands. Bill hired Captain William Dippel, who was from an old steamboat family from St. Louis, to supervise the steamboat operation; other captains were J.B. Lewis, who captained the *Sarah*, and Captain Sanborn, of Keokuk, Iowa.

The method of transferring the freight from the railroads to the barges was ingenious. The freight was switched into the yards and then switched onto the barges by a double track incline. The steamboats and barges connected to the incline by an apron; chains and anchors were employed to hold the steamboats and barges securely to the apron; the aprons were shaped like cradles and fit the u-shaped barges in a solid joint. Both the inclines and the aprons could be raised or lowered, according to the waterlevel of the river.

Captain Dippel supervised the maintenance and repairs of the boats and barges; the yard superintendent looked after the repairs to the freight, locomotives and passenger cars as they came through Angola. Additionally, there was a team of divers with complete diving equipment who maintained the inclines, did the pile-driving and repaired the inclines. These teams operated so smoothly that Bill could safely leave that part of his operations to them.

When the Vicksburg, Shreveport and Pacific Railroad abandoned their ferry transfer at Vicksburg, Bill purchased the steamer *Kellogg* (named after William Pitt Kellogg, the reconstruction Louisiana governor). The *Kellogg* was an auxiliary boat, capable of carrying 19 freight cars. Because she had her own tracks to accommodate the railroad cars, she did not push barges; instead she was mainly used to pump gravel from the river bottom and she had pumping equipment mounted overhead out of the way of the railroad cars.

There were constant challenges to Bill's ingenuity in running his railroads. Tunica Hill was just one of these challenges. The hill was composed of volcanic ash which constantly slid over and against the track, pushing the track toward the river, at one time the track had been moved 37 feet off center, which caused expensive maintenance to be required. One wet spring after heavy rains, Bill attempted to widen the cut through Tunica Hill. A large cave-in took place that covered the entire work crew, locomotives, and dump cars with the wet ash. That took several months to dig out.

Tunica Hill had a grade of four tenths of one percent which meant that full tonage trains required double locomotives to make the grade; this was especially challenging during the wheat rush from Kansas in 1912 and 1913.

Another challenge was the maintenance of the bridge across Bayou Sara Creek. The approaches and steel spans required 95 foot pilings; heavy rains frequently and unpredictably would wash logs and trees into the spans, endangering oncoming trains.

And there is the story of the unknown hero of Bayou Sara Bridge: the bridge had washed out, and as a train approached from the south, a "knight of the road" who had camped nearby, crawled over the swinging bridge of ties and rails to flag the train. The train stopped in time but the hero disappeared into the pines before he could be rewarded, never to be seen again.

Bill and Sarah enjoyed their life on the trains and boats. On Friday evenings they'd catch Train Two at the Canal Street Station, depart at 5:00 and arrive by 10:00 at Angola, 130 miles away. Then they'd board the *Sarah*, escape the heat of New Orleans and enjoy the cool breezes on the river as they rode back and forth to Naples.

The *Sarah* was a floating home for them; the staterooms, galley, pantry, living room and office were on the hurricane deck. The rooms measured 6 x 12 feet, with private commodes between some of the rooms. Each room had a berth, tin basin and pitchers; screens and windows protected the occupants from the swarms of mosquitos. And in the salon there was a picture of Sarah, as there was a picture of Bill aboard the *William*.

They enjoyed entertaining aboard the *Sarah*. One of their favorite guests was H. L. Hunt of Shreveport, and inveterate gambler. Out of respect to Captain Dippel, Hunt confined himself to playing friendly games of 7-up and solitaire. Later, Hunt was to bring in the giant east Texas oil field along the Louisiana Railway and Navigation Company of Texas; his sons would discover the huge Black Lake oil field some ten years later.

Bill and Sarah also loved playing with little Gordon and Dorothy Dippel, who were like grandchildren to them. Bill taught Dorothy German and told her stories and nursery rhymes in German, continuing in English when she became confused.

They listened to the victrola, playing the songs that Bill loved and that he named towns after: "Aloha Oe," "La Paloma," and "Sorrento." Often the *Sarah* engineer would play the accordian for them while Sarah darned clothes and socks. Sarah enjoyed cooking, so she'd furlough the cooks and crew while she cooked for everyone; one of her specialties was spaghetti and meatballs.

Often, they'd take little Dorothy to visit them in New Orleans. When Bill bought the railroad from Shreveport to Dallas, he showed Dorothy the one million dollars in cash that had been used to purchase it.

Bill and Sarah received visits from the gorgeous Arabella Huntington, who was buying Chantilly Lace of New Orleans, and was becoming the world's foremost collector of eighteenth and nineteenth century furnishings and art, including the majestic Gainsburough "Blue Boy". They visited her New York mansions which she had so lavishly remodelled and embellished with murals. Some of her murals have been transplanted to Yale University.

Her Italian Palozzo on Nob Hill had recently been destroyed by the San Francisco earthquake and then given to San Francisco as Huntington Park. Bill and Sarah had been guests there during a visit to the San Francisco wire factory. Arabella's neighbors were the other members of the Big Four Builders of California: Stanford (Stanford University), Crocker (Crocker Bank), and Mary Hopkins (Top fo the Mark Hotel).

Across the street from the Hopkins' mansion was the residence of the Silver King of the Comstock Lode, James C. Fair, present site of the most beautiful hotel lobby, the Fairmont Hotel.

The Huntingtons had purchased large parts of the Los Angeles area when it was an arid basin with barely enough water for the native Mexicans. It looked like a bad real estate deal, but it had a harbor, a potential gateway to the Orient. They then connected "Our Lady Queen of Angels" with C.P. Huntingtons' Southern Pacific and another railway from San Francisco. Bill and Sarah were hoping to accomplish the same thing in the industrially deprived Louisiana.

In Arabella's household were many mansions, including a Parisian hotel. She donated much land and made her art works available to the public. Henry Huntington, who was Collis' nephew, had redecorated an old Spanish villa of San Marino to help woo her for his own (later, successfully) and today this old Spanish villa is the Huntington Museum outside Los Angeles, the depository of most Arabella's treasures which she collected for public use. Today, the public can view the famous "Tragic Muse" painting known to the Huntington family as "Twinkie." The muse winks. They admired Arabella who spent a great effort traveling to castles and galleries and was never fooled by a forgery. She had done her homework.

THIRTEEN

The Beginning of Reforestation Urania, Louisiana

Henry Hardtner of Pineville, La., with a capital of $1000.00 began cutting yellow pine in the mid 1890's. In 1898 Bill met him on a train to New Orleans and invested $5000.00 to purchase what later became the Urania Lumber Company, 130,000 acres of the best pine lumber in the South.

Bill and Henry purchased a mill 35 miles south of Alexandria on the Missouri Pacific Railroad. They named it after the Greek muse of astronomy, Urania. (The men who worked there said it was like heaven compared to the site called Blackberry, because they'd had to live on blackberries when the mill wasn't running.) Bill and Henry began with a 3000 acre tract of virgin pine which cost less than $5.00 per acre. In eight years, due to Henry's persistence, their investment had tripled.

Bill began to fear that the pines of the South would disappear entirely. Due to the inroads on white pine by the capitalists who had "cut out and got out" and the further wholesale cutting of pine throughout Louisiana, Bill was afraid that unless selective cutting procedures were enforced, and some kind of reforestation were instituted, that entire forests would disappear. Bill also knew that most of the other lumbermen thought reforestation was useless because of the damages done by razorback hogs and forest fires. But he knew, too, that some form of reforestation had been practiced in Saxony, Germany since the 14th Century; and that there had been extensive replanting of seedlings with fire protection in Germany. So he talked it over with Henry and they agreed that they could buy cut-over lands for $1.00 an acre and begin a reseeding program. They believed that the forest could regrow within 40 years, rather than the 100 years other lumbermen believed.

Then in 1908 Bill and Henry attended Theodore Roosevelt's conference of Governors. Roosevelt's chief forester, Gifford Pinchot, of Louisiana, attended too. Louisiana took the lead in reforestation during that conference and it is believed that this began the national conservation movement.

Later, with Bill's backing, Henry became State Senator and First Chairman of the Louisiana Conservation Commission. Henry signed the first reforestation contract with the state. They established experimental plots of baby pines, although many lumbermen joked about them. The reforested lands were given tax relief by the state. Eventually other lumbermen followed and now there is a matured second growth over the entire South. And it all started at Urania, Louisiana.

Bill had written to his cousin, Paul Sippel, in Ludenshied, Germany, to come to New Orleans to "get an education in the railroad business." In August, 1913, Paul accepted Bill's invitation and was given a job as a warehouse laborer, handling broom straw, rough rice, molasses and cotton for the various ships that left port. Paul worked in

Bill's largest warehouse near the Basin Canal; he had to lay the cotton bales flat so they could be sorted for their destination. After they have been inspected, the bales were upended and marked with flags. This was hot, hard work and Paul earned $30 per week.

Then Bill promoted Paul to warehouse foreman; then as clerk in the office, Paul did the billing, checking, and applied seals to merchandise cars (14 to 18 per day) loaded for stations upriver to Angola.

Next Bill borrowed Paul and put him in charge of his oil interests in the Critchton-Lenzberg oil field, one of the largest in the South. Paul began working on the oil rigs, then progressed to field superintendent in charge of drilling, production and shipping which was done by tank car. Bill had his own refinery in the field and used the output to run his 60 locomotives.

Lenzburg and Critchton grew into two ambitious towns during this oil boom. Bill built two combination passenger and freight stations and many miles of side track to service the development, which he considered to be the brightest jewel of the Edenborn line. However, the town of Lenzburg has disappeared since that time into the Red River.

During the Lenzburg oil development, Bill and Sarah enjoyed the Bernstein house in Coushatta and were visited often by young Charlie Gates who hoped that Lenzburg would become another Spindletop.

Bill and Sarah led frugal and industrious lives; they spent most of their time either on one of their trains, or on a boat, usually the *Sarah*, or visiting one of their homes with the beloved Dippel children.

Bill's activities in the lumber business continued; he purchased the cut over lands of the Big Pine Lumber Company. Because most of the 60,000 acres were in Grant Parish, Bill renamed the company; Colfax Lumber and Creosote Company, after Schuyler Colfax, President Grant's Vice-President.

At one time Bill owned over a million acres of Louisiana timberlands; he would establish lumbermills and towns and name them after friends and acquaintances. One of these towns was Conrad in the Black Lake area where Bill and Sarah would enjoy fishing and vacationing even while Bill supervised the lumber business there. Conrad had been established to harvest the yellow pine there; and once the pine had been cut, Bill attempted to replant the forest with seedlings because cut over land had little value for farming or cattle raising. But these reseeding projects faced almost impossible obstacles: forest fires, and the wild hogs that had escaped captivity would eat anything, including the tender seedling trees.

In addition to reforestation and conservation, Bill was interested in preserving the giant cypress forests for later generations. He was particularly interested in the cypress stand at Coochie Brake (now a government preserve) where the trees were as tall and old as the redwoods. Stories told that these trees were so huge that a thousand acres could produce 60,000,000 board feet of lumber.

And even while he supervised his lumber business, Bill was active on his railroads. A story about Bill and his activities on a train: he'd helped to pull some injured men from a derailed train and when he checked into a hotel, dusty and tired, the clerk asked if he were a conductor or engineer. Bill had replied: "I own the railroad but the engineers and conductors make more money than I do." He was referring to the fact that the claims and deficits on the transfer operations of his railroad/steamboat lines cost him $100,000 to $600,000 (that would be over six million dollars today). Since he didn't have shareholders, Bill carried those deficits himself.

There were other significant challenges that Bill had to handle. Martin Behrmann, the Mayor of New Orleans, came to Bill at the beginning of World War I to tell Bill that some of Bill's political competitors were working to shift public opinion against Bill and his various enterprises. Bill had established the town of Wilhelm north of Baton Rouge; now that the War had begun, Bill's competitors hinted broadly and publicly that Bill's sympathies were with the Kaiser. To forestall such a political move, Bill renamed the town Tunica; this proved wise, for the 5,000 who worked for Bill approved.

Then there were the floods. One of the greatest floods of the Mississippi and Atchafalaya Basin occurred in 1912. The flooded areas extended from Hessmer to Baton Rouge and New Orleans.

Bill's levees in the Bayou De Glaize area could not contain the flooding; miles of track for the Louisiana Railway and Navigation Company, Texas and Pacific, and Southern Pacific Railroads were washed out.

Bill and Sarah rushed to the flood scene and stayed for two months to help restore the rail lines. Transportation was completely halted and the transfer boats were requisitioned by the government to perform relief services for the people and their possessions from Hessmer to Baton Rouge. Thousands of people were transported by the Angola steamers from the flood district to special camps at Hessmer and Mansura, where they lived for three or four months. The railroad supplied food, clothing and medicine for the refugees.

When the governor came out on a special train to Hessmer, he was astonished by the hundreds of miles of flooded terrain; the waters were 20 feet deep in places and the steamboats floated above the telegraph lines!

The other major flood occurred in 1922, broke through the levee at the Angola prison farm and trapped 200 inmates on the second floor of the sugar mill. Bill asked the Standard Oil refinery at Baton Rouge to send some of their steel hulled boats to the prison to rescue the prisoners; however, the Standard Oil pilots did not know the waters and didn't want to take the risks. So Bill asked his own Captain Kelly King if he'd risk the wooden hulled *J.B. Lewis* in a rescue attempt. King agreed and the wooden hulled *J.B. Lewis* rescued 200 prisoners from the rapidly rising waters.

All during this time Bill had been sharing the Union Depot in New Orleans with two other railroads. When he asked the Mayor, Martin Behrmann for permission to build his own terminal for the Louisiana Railway and Navigation Company, the Mayor refused. However, when Behrmann lost the next election for mayor, in 1920, Bill got the necessary permission from the new city government and built his own terminal on South Rampart Street, near the site of the present Super Dome.

One of the last major problems that Bill tackled was the boat transfer through the Red River at Turnbull Island. They couldn't bridge the trains there because floods submerged most of the island frequently. Bill decided to dig a ditch through the narrow part of the island, hoping that the river would deepen and enlarge the ditch to a navigable channel. There was some fear that the Mississippi would revert to her old route if this channel were deep enough; however, due to some well constructed dams in the Atchafalaya (built in 1880-1882) the Mississippi was constrained. But Bill's idea was not a good one. After a lot of digging with mules and scoops the ditch was completed and it was expected that the next time the river rose it would eat out a channel. Instead it only created a shallow slough.

Bill, against the advice of Paul Sippel and Captain Dippel, decided to run the *William* through the channel. He got behind the wheel himself and applied power to the paddle wheels. He started through the channel but small trees stopped the boat, breaking guard rails and smashing against the cabin. Captain Dippel said, "We're tearing her up." But Bill, stubborn, said, "They're my boats, just you get the carpenters to fix them."

This was one of Bill's mistakes; later the Corps of Engineers made him fill in the channel.

The last major bridge that Bill supervised was at Simmesport where he built a transfer line across the Atchafalaya. He was 78 years old and supervised every detail of this bridge.

It was here that the steamboat, the *William Pitt Kellogg* sank. She had been heavily loaded and as she headed toward a pier in the middle of the river, a forward hatch began to take in water. Bill and Paul who were aboard, tried to close the hatch, but had to leave the boat themselves. Bill, soaking wet, had to be helped aboard a skiff, where they watched the steamboat sink.

EMDEN, Louisiana, near Winnfield

Bill had bought a farm near Winnfield, Louisiana, and decided to do his agricultural research there. He named the farm Emden after the place where the Ems River empties into the North Sea. He befriended Seaman Knapp of Lake Charles from Iowa State University and put Knapp's ten commandments to work on his own projects: "well drained, well pulverized seed bed, seed of best variety, plenty of growing space, intense tillage during growing season, use of fertilizer, crop rotation, better implements and more horsepower, utilization of waste products and idle lands with livestock, production of all food, and a complete accounting system."

Bill supervised research projects such as protection of cotton from boll weavils, developing silkworm growth, and even a peanut farm, for which he received encouragement from George Washington Carver. It was at Emden, when he was seventy three, that Bill invented, and received his last patent #1,351,629, for the apparatus that distilled pine oils for turpentine. The pine stumps would be ground into chips which were boiled in a huge tank, and the tar and resins were distilled off. Housed in a two story building, this appartus made turpentine more efficiently than any other previous method.

Much of the land that Sarah and Bill owned east of Emden was purchased from descendants of the Cane River people, who themselves were the descendants of freed slaves or "Creoles of Color." About 1800 they had begun to settle a wet area south of Natchitoches; this was lowland between the Cane River and the Red River, and had been owned by a man named Bossier. The land was either given to them, sold to them, or obtained by squatter's rights; and by the time of the Civil War, the Cane River people had an empire of their own, based on slavery and cotton trade. Although they had been loyal to the South during the war, after the war the Cane River people were treated with contempt by both sides. However, they kept their great plantations and established settlements on Bill's railroad such as Grappes Bluff and Clothierville, Louisiana.

Bill spent his last days at Emden, still planning new projects. He and Sarah planned to make a garden spot of Emden and to plant lilies and hyacinths there. But one afternoon in the middle of these plans, Bill complained of chest pains and said that he was tired. He died quietly,

LOUISIANA RAILWAY AND NAVIGATION COMPANY

EDENBORN LINE

American Express.

GENERAL OFFICES SHREVEPORT LA

WILLIAM EDENBORN President. 820 Hibernia Building New Orleans La
H B HELM. Vice-President. Shreveport La
J J TIPPIN. Auditor and Secretary.
L M McFARLIN Treasurer
E H RANDOLPH. General Counsel
E C D MARSHALL Gen Freight & Passenger Agt
W H COPPAGE. Superintendent
C C COLLEY Purchasing Agent.
T NICHOLSON. Master Mechanic
E F BUCHANAN Car Accountant
C R MEE General Roadmaster.
C E LEWIS Claim Agent
D M DODGE. Gen Agent 708 Victor Bldg Kansas City Mo
J F MURDY. Commercial Agent New Orleans La
J R JORDAN. Commercial Agent Baton Rouge La
O N NESOM. Commercial Agent Alexandria La
R L McKee Commercial Agent 307 Southern Trust Building Little Rock Ark
J E BLAINE Commercial Agent 1213 Busch Bldg Dallas Tex
C F SMITH Commercial Agent 210 Insurance Building Oklahoma City Okla
M S SUTHERLAND European Freight Agent 76 Newgate Street London E C Eng
T C FREW Commercial Agent Shreveport La
O B BATTON City Ticket Agent
GEORGE HIXON. Commercial Agent Chattanooga Tenn

Electric-lighted Pullman-Buffet Drawing-room Sleeping Car Service

BETWEEN

New Orleans, Baton Rouge, Bayou Sara, Alexandria and Shreveport On Trains Nos. 1 and 2

CONNECTIONS

Shreveport. La—With Houston & Shreveport R R Kansas City Southern Ry. Louisiana & Arkansas Ry. Missouri. Kansas & Texa Ry of Texas Texas & Pacific Ry St Louis Southwestern Ry. and Vicksburg. Shreveport & Pacific Ry

Hagen. La —With Louisiana & North West R R

Winnfield. La —With Louisiana & Arkansas Ry. Tremont & Gulf R.R., and Chicago. Rock Island & Pacific Ry (Choctaw District)

Alexandria. La —With Missouri Pacific R R. Louisiana & Arkansas Ry. Texas & Pacific Ry. C R I &P Ry and Morgan's Louisiana & Texas R.R & S.S Co.

Scotland. La —With Yazoo & Mississippi Valley R R

Baton Rouge. La —With New Orleans. Texas & Mexico R R. Yazoo & Mississippi Valley R R

New Orleans. La —With Illinois Central R R. Louisiana Southern Ry. Louisville & Nashville R R Morgan-s Louisiana & Texas R R & S S Co. New Orleans Terminal Co New Orleans Southern & Grand Isle Ry. New Orleans Great Northern R R. New Orleans & Northeastern R R. New Orleans Texas & Mexico R R. Texas & Pacific Ry., and Yazoo & Mississippi Valley R R

WINNFIELD BRANCH.

......	No. 20	Mls.	March 14.	Mls	No. 19	......
			LEAVE ARRIVE			
	*6 50 A M				7 15 P M	
	- -		Red River Bridge		- - -	
	8 51 A M		**Coushatta**		6 07 P M	
	10 16 A M		arr .... **Aloha** .... lve		6 20 P M	
	*3 20 P M	0	lve ... **Aloha** δ ... arr	27 6	9 30 A M	
	3 50 P M	8 2	Verda δ	19 4	8 25 A M	
	- -	10 1	Lofton	17 5	- -	
	4 11 P M	11 8	Emdem	15 4	7 59 A M	
	4 60 P M	16 3	Atlanta δ	11 3	7 35 A M	
	4 59 P M	10 0	Whitford	8 6	7 25 A M	
	6 05	21 2	Alonzo	6 4	7 20 A M	
	6 30 P M	27 6	+ **Winnfield** δ	0	*7 00 A M	
			ARRIVE LEAVE			

EXPLANATION OF SIGNS

Trains marked * mn daily. 1 Meals + Coupon stations. δ Telegraph stations. STANDARD—Central time

15	5	1	3	Mls	March 14	Mls	4	2	6	16
		P M	A M	0	lv ·Shreveport δ ar	306 5	P M	A M		
		*500	*6 50		Union Station		7 16	7 30		
		5 21	7 16	5 4	Blenheim	301 1	6 61	7 04		
		5 29	7 27	9 3	Curtis	297 2	6 40	6 50		
		6 40	7 37	14 2	Taylortown δ	292 3	6 30	6 40		
		5 61	7 46	18 2	Elm Grove	268 3	6 20	6 30		
		6 56	7 51	20 4	McDade	286 1	6 16	6 22		
		6 00	7 55	21 5	Atkins δ	285 0	6 11	6 20		
		6 05	8 00	2 30	Poole	283 9	6 06	6 14		
		6 14	8 08	27 5	Ninock	279 0	5 54	6 04		
		6 20	8 15	30 3	Loggy Bayou	270 2	5 47	5 57		
		6 28	8 22	33 1	East Point δ	273 4	5 40	5 50		
		6 37	8 32	36 2	Crichton	270 3	6 30	5 41		
		6 39	8 34	37 3	Lenzburg	268 7	5 28	5 39		
		6 46	8 40	32 6	Carroll	266 9	5 18	5 31		
		6 57	8 51	45 4	Coushatta δ	261 1	5 07	5 19		
		7 10	9 02	50 9	Pirmont	255 6	4 54	5 08		
		7 20	9 11	56 1	Grappe's Bluff	250 4	4 44	4 58		
		7 38	9 24	62 3	Campti δ	244 2	4 31	4 45		
		7 42	9 32	65 7	Hagen δ	240 8	4 23	4 38		
		7 51	9 41	69 2	Clarence δ	237 3	4 14	4 29		
		7 54	9 45	71 0	Irma	235 5	4 09	4 25		
		7 59	9 49	72 0	Luella	233 6	4 05	4 20		
		8 05	9 54	75 0	St Maurice	231 5	4 00	4 14		
		8 14	10 03	78 2	Crews	228 3	3 53	4 06		
		8 24	10 13	82 9	Montgomery δ	223 6	8 40	3 54		
		8 42	10 39	92 2	Aloha δ	214 3	3 20	3 36		
		8 66	10 51	98 0	Colfax δ	208 5	3 01	3 20		
		- -	10 54	99 1	McNeeley	206 9	2 68	- -		
		- -	10 38	101 7	Ravencamp	204 8	2 53	- -		
		9 07	11 04	103 4	Kateland	203 1	2 48	3 08		
		- -	11 10	107 0	Werdohl	199 5	2 41	-		
		9 19	11 16	109 2	Meade	197 3	2 35	2 59		
		- -	11 22	11 29	Tyrawley	193 6	2 27	- -		
		- -	11 28	11 47	Barrett	191 8	2 23	- -		
		8 47	11 44	121 2	Pineville δ	185 3	2 10	2 29		
P M		9 58	11 59	12 26	ar · Alexandria δ lv	183 9	*2 00	2 19		
*3 15		10 08	A M	122 6	lve Alexandria arr	181 9	P M	2 09		12 25
3 30				125 7	Arno	180 8		- -		12 00
3 65		10 31		131 7	Latanier	174 8		1 42		11 30
4 08		10 38		135 5	Richland	171 0		1 33		11 10
4 21		10 45		137 6	Magda	168 9		1 27		11 00
4 35		10 52		141 1	Bijou δ	165 4		1 19		10 45
4 53		11 00		145 3	Belledeau	161 2		1 08		10 25
6 10		11 11		140 3	Hessmer δ	157 2		1 00		10 05
5 35		11 19		153 5	· Mansura δ	153 0		12 52		9 40
6 00		11 29		157 0	Moreauvile δ	148 6		12 42		9 15
6 15		- -		161 1	Willard	145 4	- -			9 02
6 50		11 42		163 6	Bordlelonville δ	142 9		12 29		8 50
6 54		- -		166 4	Zimmer	140 1		- -		8 37
6 58		- -		167 1	Rexmere	179 4		- -		8 34
7 12		1* 57		169 0	Kleinwood	137 5		12 17		8 27
7 25		11 59		172 6	Sarto	135 9		12 13		8 17
7 39		- -		172 1	Cortes	134 4		- -		8 12
7 55		12 20		175 0	Naples δ	131 5		*11 59		*8 00
P M	P M	A M			(Angola Transfer Co)			P M	Noon	A M
	*200	2 55		17 76	Angola δ	128.0		10 00	12 01	
	2 13	3 09		183 1	Wilheim δ	123 4		9 46	11 47	
	2 18	- -		185 7	Brandom	120.8		- -	11 41	
	2 26	3 23		188 4	Plettenberg	118 1		8 33	11 33	
	2 40	3 48		108 3	· Bayou Sara δ	108.2		8 08	11 00	
	2 59	- -		202 1	Soest	104 4		- -	10 59	
	3 05	- -		205 0	Mannheim	101 5		- -	10 55	
	3 08	- -		206 3	Paloma	100.2		- -	10 50	
	3 13	4 15		208 7	Port Hudson	97 8		8 43	10 43	
	3 17	4 18		310 2	Bohn δ	96.3		8 39	10 39	
	3 26	4 28		213 4	Irene	92 7		8 30	10 30	
	3 42	4 46		220 6	Scotland	85.9		8 13	10 14	
	4 00	5 05		326 1	· Baton Rouge δ	80 4		8 00	10 00	
	4 13	5 18		232 2	Essen	74 3		7 42	9 42	
	4 18	5 23		234 2	Siegen	72 3		7 37	9 30	
	4 24	5 28		237 3	Kleinpeter	69 2		7 32	9 31	
	4 29	5 33		239 6	Bullion δ	60 9		7 27	9 26	
	4 35	5 37		212 0	Nettie	64 5		7 22	9 21	
	4 40	5 45		244 7	Witten	6 18		7 17	9 16	
	4 46	5 49		246 4	Gonzales δ	59 7		7 12	9 11	
	4 51	5 55		250 4	Brittany	56 1		7 06	9 03	
	4 58	6 02		252 4	Sorento δ	54 1		7 01	8 58	
	5 04	- -		255 3	Barmen	51 2		- -	8 53	
	6 03	6 13		257 0	McElroy	40 5		6 50	8 48	
	6 23	6 25		263 4	Waldeck	43 1		6 53	8 30	
	5 28	6 30		265 5	Gramercy δ	41 1		6 34	8 31	
	6 37	6 34		169 7	Garyville δ	36 8		6 26	8 23	
	6 41	- -		271 6	Terrehaute	34 9		- -	8 19	
	5 44	6 46		272 9	Reserve	33 6		6 20	8 16	
	6 51	6 53		276 6	Montegut δ	29 9		6 12	8 08	
	6 54	6 56		178 1	Ory	28 4		6 08	8 05	
	6 57	6 59		279 2	Elvina	27 3		6 05	8 02	
	6 02	7 03		281 1	Alcazar	25 4		6 02	7 57	
	6 26	7 25		291 5	Frellsco	15 0		6 33	7 38	
	6 32	7 31		204 3	Kenner δ	12 2		6 33	7 31	
	7 05	8 05		306 5	arr · New Orleans δ lve	0		*6 00	*6 55	
	P M	A M			(Canal St Station)			P M	A M	
					(New Orleans Terminal Co)					

probably still planning new projects.

AFTERWORD

William Edenborn was not a notorious figure. Because he was never involved in scandal, his life, as eventful as it was, passed without publicity. He was, however, a vital figure in the life of the rail lines between St. Louis and New Orleans, and there are countless anecdotes about him.

Some of these anecdotes are about Bill and the famous people he knew. Teddy Roosevelt enjoyed hunting, fishing and camping with Bill at Black Lake, La. It was one of these trips that Roosevelt caught the choupique fish. These fish are the only surviving descendants of dinosaurs, extremely ferocious and because they are equipped with air-breathing apparatus, able to exist on land. Roosevelt was immensely pleased with this catch and held Bill to his open invitation to return often to the fishing retreat.

Another anecdote involved Bill's nephew, Paul Sippel and Huey P. Long. Long, at that time, was the railroad commissioner of Louisiana, and required that Paul submit written reports about Bill's railroads. Too often, Long would return the reports to Paul for revision. Finally, Paul told Long, "Huey, if I have to keep making you this report, I am going to beat the shit out of you." And Long believed him, for he said: "Paul, you don't have to make *that* report anymore."

It was the same Paul Sippel that dared to tease Bill Edenborn about his stinginess. Although Bill was a multimillionaire, he was not only frugal he was stingy. He spent less than $200. per year year on himself and his family's personal needs. So stingy was he, that he would walk along the railroad tracks with a tin can and gather the spikes and nails that he found fallen alongside, and save those for later use! It has been reported earlier that he and his wife when entertaining such people as Teddy Roosevelt, John D. Rockefeller, Bet-A-Million Gates, or August Busch, that they would serve home made sandwiches, play their victrola and enterain in the simplest of fashions. Once, Paul Sippel seeing Bill wearing an ancient blue serge suit that has acquired a sheen of its own, said, "Before you sit down, take the pants off so you won't shine the suit."

On an earlier occasion, Paul and Bill had been walking along a newly dug ditch, and Paul had been wearing a new suit. Bill, concerned about the new suit had cautioned Paul to stay away from the edge. When Paul slipped and fell into the ditch and got the new suit wet, Bill had said, "That's a good experience for you."

Bill Edenborn is remembered today by many of the people he employed, and there were thousands of them, for he invested his own money in the railroads that he built, and the towns that he established along the railroads were populated by people that he hired in all capacities from railroad engineers to conductors to track layers. Additionally, the townspeople themselves, from small store owners to rooming house managers, all felt the influence of Edenborn industry and money in their lives. He seemed to be comfortable with everyone, entertained everyone he met either aboard the Sarah (which his wife had turned into a floating garden with her potted ferns and other favorite plants) or aboard their private railroad car. Bill, while he became even more penurious as he aged (one example of that is: he and Sarah owned a "shotgun" house, that is a house that could be divided and sublet, and he always had that other half of the house rented out or sold to someone else!) was an extremely generous and forgiving employer. Stories are told of how often a man might be fired by a foreman for drinking or laxity or even theft, and if the fired man went to Bill, would often be reinstated! And, because he was

LOUISIANA RAILWAY & NAVIGATION COMPANY OF TEXAS

GENERAL OFFICES—GREENVILLE, TEXAS

General Offices—Greenville, Tex.

Mrs. SARAH EDENBORN, President.	Shreveport, La.
M. A. STAMAN, Executive Vice-President.	"
PAUL SIPPEL, Assistant to President.	"
R. R. FARMER, Vice-President.	Greenville, Tex.
R. M. MILLING, General Solicitor.	New Orleans, La.
M. MAHAN & DOHONEY, General Attorneys.	Greenville, Tex.
H. L. MONEY, Assistant to Vice-President and Purchasing Agent.	"
J. N. CAMPBELL, Traffic Manager.	"
A. H. RAGAN, General Frt. and Pas. Agent.	"
R. L. COLEMAN, Generan Agent.	Dallas, Tex.
R. D. SHROPHIRE, Jr., Commercial Agent.	Fort Worth, Tex.
E. D. HICHMAN, Commercial Agent.	Shreveport, La.
Dr. WILL CANTRELL, Chief Surgeon.	Greenville, Tex.
H. P. TWYMAN, Auditor, Secretary and Treasurer (in charge of freight claims).	"
M. M. MILLER, Assistant Treasurer.	"
M. B. THOMAS, Claim Agent.	"
O. T. GRUENE, Car Accountant.	"
W. C. BORCHERT, Chief Engineer.	"
J. W. VARLEY, Master Mechanic.	"

This line also operates dependable FAST FREIGHT SERVICE, reaching with its own rails or through connecting lines New Orleans, Baton Rouge, Alexandria, Memphis, St. Louis, Kansas City, Dallas, Fort Worth, also eastern, northern and western points.

Total Mileage, 204 Miles.

McKINNEY BRANCH.

11 Mix	Mls	January 20.	12 Mix
A M		LEAVE — ARRIVE	P M
β7 30	0	**Greenville** δ	3 05
8 30	8 0	Floyd δ	2 15
9 20	14 9	Farmersville δ	1 25
10 10	24 3	Princeton δ	12 4U
10 50	32 3	· **McKinney** δ	β1250
A M		ARRIVE — LEAVE	P M

GREENVILLE AND SHREVEPORT.

1	January 20.	Mls.	2
AM	LEAVE — ARRIVE		P M
˙1020	· **Greenville** δ	172 1	2 58
10 45	Campbell δ	161 7	2 32
11 01	Cumby δ	155 0	2 17
11 14	Brasher	148 8	2 03
11 33	· **Sulphur Springs** δ	141 0	1 47
11 46	Crush	135.1	1 30
11 52	Como δ	131 6	1 23
12 07	Pickton δ	126 4	1 10
12 45	**Winnsboro** δ	118 4	12 52
12 47	Eser	117 3	12 30
† -	Musgrove	114 5	† -
12 59	Scroggins	111 9	12 17
1 05	Newsome	109 1	12 10
1 12	Leesburg δ	105 7	12 02
1 33	· **Pittsburg** δ	99 0	11 47
1 45	Baker	94 8	11 34
1 59	Cason δ	89 1	11 19
2 12	Daingerfield δ	83 3	11 04
2 30	Hughes Springs δ	77 0	10 50
2 45	Patman	72 0	10 35
2 56	Avinger δ	68 1	10 26
† -	Orrs	63 5	† -
3 11	Lassater δ	61 7	10 12
† -	Sarber	58 7	† -
† -	Kelleyville	56 5	† -
3 40	**Jefferson** δ	49 7	9 46
3 53	Baldwin	42 0	9 24
4 10	Karnack δ	37 2	9 13
† -	Fitzpatrick	35 2	† -
4 20	Leigh δ	32 4	8 59
† -	Lotta	30 0	† -
4 30	Lemo	28 4	8 50
† -	Blocker	26 6	† -
4 47	Waskom δ	22 5	8 37
6 05	· **Shreveport** δ	0	˙7 25
P M	ARRIVE — LEAVE		A M

EXPLANATION OF SIGNS.

Trains marked ˙ run daily. £ daily, except Sunday: † flag stop

· Coupon stations

δ Telegraph stations

comfortable with everyone, he made friends with people that might have been regarded with caution.

Among these friends was a lumberman who had a reputation for troublemaking, he was called in the slang of those days, a bummer, he had caused a lot of fights in Atlanta, Louisiana on Saturday nights and had, in general, raised hell in Atlanta. He was called "Hum" Dinger, and after several arrests had been warned out of town by one of the local judges. As Hum Dinger told Bill, he'd been arrested for disturbing the peace. And Hum Dinger had said to the judge; "Your Honor, there ain't no peace in Atlanta on Saturday nights!"

Another of these friends was "Leadbelly" Ledbetter, a black who had worked for Bill as a "spike mauler" in one of Bill's lines. Bill and Sarah went to visit Leadbelly while he was interred at the Louisiana State Prison for "knocking someone over the head" and seriously damaging him. Leadbelly, a songwritter, had written "John Henry" as well as several other work songs he had heard during his days on the chain gangs. Leadbelly wrote, with John Lomax, the Jazz historian, the song, "Goodnight Irene," about Bill's town of Irene. These were the days when the chain gangs were loaned out to work for various raillines. While Leadbelly did not object to this practice, Bill did, and lobbied against it. There was, also, an anecdote about how John D. Rockefeller, a fellow director of U.S. Steel, liked to ride Bill's Railroad while his Standard Oil Refinery was being established at Baton Rouge on Bill's railroad. He'd ride on a flat car and enjoy the scenery; later one of the Public Service Commissioners insisted that these flat cars be roofed, but not before J.D. Rockefeller had enjoyed several rides on this "commuter train."

If there were humorous tales about Bill and his many friends there was also at least one about the many rundown railroads that Bill bought. In 1923, Bill decided to buy the route from Dallas to Shreveport, which was broken down in the extreme. The only time the rails were level was

Sarah Edenborn - Railroad President

when one rail was going up and the other was coming down. Bill and Sarah purchased this line because they believed in the future of the Shreveport area and moved their home to Shreveport.

As mentioned throughout the book, Bill loved to name towns after friends and acquaintances. One could track his friendships, and business dealings, by the towns he named and established. Because his wealth was well known, Bill had to be circuitous in some of his land purchases in order to avoid artificially raised prices. To avoid these prices, he'd ask relatives like Ed and Alonzo Drain of St. Louis to buy land for him; thus the town of Alonzo was named. Bill named Randolph after his first attorney and landman; he named Nicholson, Louisianna after a mechanic who'd worked for him, The Winnfield Line "Stops" similarly named; Bill was especially fond of telling stories of how the line to Winnfield was alight for twenty-four hours a day at the height of its lumber cutting days when thousands of board feet of virgin timber were processed through

Winnfield. Luella was named after the daughter of a plantation owner that Bill and Sarah visited often.

Bill loved to name towns after favorite songs: Aloha, after a Hawaiian song, Sorrento, after still another.

He named towns after places in Germany and the Ruhr Valley: Bingen, Lenzburg and Hagen, Wilhelm, later renamed Tunica, as well as Emden, on the Ems River, in Germany.

Of course, there was Nettie, after Antoinette, as well as Edenborn, La. He named towns too, for business acquaintances: Barnam, after A. Gonzales, LA alderman; the "ammonia" king of St. Louis, Edward Mallinckrodt.

And then there were the guests and visitors that Bill and Sarah entertained aboard the *Sarah* and aboard their private railroad car. These guests included not only Theodore Roosevelt, but his nephew, Franklin Delano Roosevelt. Such fabulous names as J.P. Morgan, Charlie Gates, Elbert Gary, H.L. Hunt, Charlie Steele (the 1892 partner of J.P. Morgan), G.H. Walker, Adolphus Busch, as well as but at different times, William Anheuser; such men as E.H.R. Green, the son of Hetty Green, who himself had a fortune and a research farm like Bill's; and Gifford Pinchot, the Chief Forester of the United States, all these men ate sandwiches and listened to the victrola with Sarah and Bill. Governors of Louisiana; Colonel Sam Jones and Governor Blanchard, so did mayors of New Orleans come aboard; Martin Behrmann and Paul De Capdeville. Harvey Couch, utility magnate, came aboard, as did Henry Hardtner, "the father of reforestation;" and Ed and Alonso Drain, his brothers-in-law and relative Johnny Burns, the master mechanic of the Missouri-Pacific; and Michael Kelly, the roadmaster who laid the Missouri-Pacific track to Washington, Mo. And John Hay who became Secretary of State, and whose grandson-in-law, was Stuart Symington, added to the illustrious and always colorful lists of guests. McNeely near Colfax, a business partner; Bullion for the man from whom he had bought land; Harahan, an Illinois Central executive; Conrad, a man who ran a sawmill on Black Lake and who was from St. Louis; Lofton after the Lofton brothers; Zimmerman after the man who ran the St. Louis Iron and Machine Works; plus many many others.

Books used for Reference

Anna Burns; A History of the Louisiana Forestery Commission

Steel Titan, The Life of Charles N. Schwab; **Robert Hessen, New York. Oxford University Press 1975.**

Steel Wire In America; **Kenneth B. Lewis, The Wire Association. Branford, Connecticut.**

Genesis of the American Steel and Wire Company of New Jersey; **Industrial Museum of the American Steel and Wire Company, Worcester, Massachussets.**

The Wire that Fenced the West; **Henry D. and Francis T. McCallum, University of Oklahoma Press, Norman, Ok..**

Diaries of William Edenborn, Paul Sippel and Fred Clifford; **United States Steel Annual Reports (1902 - 1909), University of Georgia Library.**

Henry Hardtner (Forests and People Magazine, fourth quarter, 1978)
Anna Burns

Railroads of America; **J.F. Stover**

Samuel. Florman, The Existential Pleasures of Engineering; **St. Martin's Press, New York, 1976.**

Frank Donovan; River Boats of America; **Thomas Y. Crowell Company, New York, 1966.**

"Bet a Million Gates" **by Lloyd Wendy and Herman Kogan; Arno Press (still in print).**

Diaries of Dorothy Drain Coleman.

ILLUSTRATIONS

INDEX

T

U

V

W

X

Y

Z